VOLUME EIGHT

ADVANCES IN CHEMICAL POLLUTION, ENVIRONMENTAL MANAGEMENT AND PROTECTION

Urban Soil and Water Degradation

VOLUME EIGHT

ADVANCES IN CHEMICAL POLLUTION, ENVIRONMENTAL MANAGEMENT AND PROTECTION

Urban Soil and Water Degradation

Edited by

PAULO PEREIRA

Environmental Management Laboratory, Mykolas Romeris University, Vilnius, Lithuania

CARLA SOFIA SANTOS FERREIRA

Department of Physical Geography, Stockholm University and Bolin Centre for Climate Research, Stockholm, Sweden; Navarino Environmental Observatory, Messenia, Greece; Research Center of Natural Resources, Environment and Society, Polytechnic Institute of Coimbra, Coimbra School of Agriculture, Coimbra, Portugal

ELSEVIER

ACADEMIC PRESS

An imprint of Elsevier

Academic Press is an imprint of Elsevier
50 Hampshire Street, 5th Floor, Cambridge, MA 02139, United States
525 B Street, Suite 1650, San Diego, CA 92101, United States
The Boulevard, Langford Lane, Kidlington, Oxford OX5 1GB, United Kingdom
125 London Wall, London, EC2Y 5AS, United Kingdom

First edition 2022

ISBN: 978-0-12-820180-0
ISSN: 2468-9289

For information on all Academic Press publications
visit our website at https://www.elsevier.com/books-and-journals

Publisher: Zoe Kruze
Acquisitions Editor: Jason Mitchell
Developmental Editor: Jhon Michael Peñano
Production Project Manager: Sudharshini Renganathan
Cover Designer: Vicky Pearson Esser

Typeset by STRAIVE, India

Contents

Contributors

B. Ayça Ataç
TU Dortmund University, Dortmund, Germany

Damia Barcelo
Catalan Institute for Water Research (ICRA-CERCA), Girona, Catalonia, Spain

Leonardo Bianchini
Department DAFNE, University of Tuscia, Viterbo, Italy

Katažyna Bogdzevič
Environmental Management Laboratory, Mykolas Romeris University, Vilnius, Lithuania

Rosa Coluzzi
Institute of Methodologies for Environmental Analysis - National Research Council of Italy (IMAA-CNR), Tito Scalo, PZ, Italy

Manob Das
Department of Geography, University of Gour Banga, Malda, West Bengal, India

Valerio Di Stefano
Department DAFNE, University of Tuscia, Viterbo, Italy

António C. Duarte
School of Agriculture/Polytechnic Institute of Castelo Branco, Branco, Portugal

Paolo Fabbri
Dipartimento di Geoscienze, Università di Padova, Padova, Italy

Carla S.S. Ferreira
Department of Physical Geography and Bolin Centre for Climate Research, Stockholm University, Stockholm, Sweden; Navarino Environmental Observatory, Messenia, Greece; Research Center of Natural Resources, Environment and Society, Polytechnic Institute of Coimbra, Coimbra School of Agriculture, Coimbra, Portugal

Francesco Gallucci
Consiglio per la Ricerca in agricoltura e l'analisi dell'Economia Agraria (CREA), Monterotondo, Italy

Navid Ghajarnia
Department of Physical Geography and Bolin Centre for Climate Research, Stockholm University, Stockholm, Sweden

Thomas Hartmann
TU Dortmund University, Dortmund, Germany

Vito Imbrenda
Institute of Methodologies for Environmental Analysis - National Research Council of Italy (IMAA-CNR), Tito Scalo, PZ, Italy

Miguel Inácio
Environmental Management Laboratory, Mykolas Romeris University, Vilnius, Lithuania

Zahra Kalantari
Department of Physical Geography and Bolin Centre for Climate Research, Stockholm University; Navarino Environmental Observatory, Messenia, Greece; Department of Sustainable Development, Environmental Science and Engineering (SEED), KTH Royal Institute of Technology, Stockholm, Sweden

Marius Kalinauskas
Environmental Management Laboratory, Mykolas Romeris University, Vilnius, Lithuania

Marijana Kapovic-Solomun
University of Banja Luka, Faculty of Forestry, Department of Forest Ecology, Banja Luka, Republic of Srpska, Bosnia & Herzegovina

Milica Kasanin-Grubin
Institute of Chemistry, Technology and Metallurgy, University of Belgrade, Belgrade, Serbia

Riad Lilipaly
Department of Physical Geography and Bolin Centre for Climate Research, Stockholm University, Stockholm, Sweden

Enrico Maria Mosconi
Department of Economics, Engineering, Society and Business Organization (DEIM), University of Tuscia, Viterbo, Italy

Pietro Daniel Omodeo
Dipartimento di Filosofia e Beni Culturali, Università Ca' Foscari, Venezia, Italy

Paulo Pereira
Environmental Management Laboratory, Mykolas Romeris University, Vilnius, Lithuania

Francesca Perrone
Department of Planning, Design, and Technology of Architecture (PDTA), Faculty of Architecture, Sapienza University of Rome, Rome, Italy

Michele Pezzagno
Department of Civil, Environmental, Architectural Engineering and Mathematics, DICATAM, University of Brescia, Brescia, Italy

Luca Salvati
Department of Economics and Law, University of Macerata, Macerata; Department of Methods and Models for Economics, Territory and Finance, Faculty of Economics (MEMOTEF), Sapienza University of Rome, Rome, Italy

Samaneh Seifollahi-Aghmiuni
Department of Physical Geography and Bolin Centre for Climate Research, Stockholm University, Stockholm, Sweden

Sebastiano Trevisani
Dipartimento di Culture del Progetto, Università IUAV di Venezia, Venezia, Italy

Farzad Vahidi Mayamey
Department of Physical Geography and Bolin Centre for Climate Research, Stockholm University, Stockholm, Sweden; Department of Civil, Chemical, Environmental, and Materials Engineering (DICAM), University of Bologna, Bologna, Italy

Series editor's preface

I am delighted to introduce the second volume in this series edited by a colleague and friend of mine, Paulo Pereira. Volume 8 is titled *Urban Soil and Water Degradation*. It is complementary to Volume 4 in this series, also edited by him, *Soil Degradation, Restoration and Management in a Global Change Context*. An introduction to the present Volume 8 was in a way advanced in Chapter 6 of the previous Volume 4 on urban areas. Indeed, management of urban areas will play a prominent role in the coming years due to rapid global urbanization.

One of the problems identified in this future scenario is water scarcity—where demand exceeds availability—and it will affect the health and wellbeing of urban residents, urban environmental quality, and socio-economic development. At present, many megacities of the world's urban populations face water scarcity. Population growth, urbanization, and socio-economic development are expected to increase urban water demand by 60%–80% over the next decades. By 2050, nearly half of the global urban population (1.7–2.4 billion) will live in water-scarce regions, with about one quarter concentrated in India, and more than 15 (>60%) global mega-cities are expected to face water scarcity. Simultaneously, climate change will affect water availability, with an increased number of extreme events such as floods. Most importantly, urban water scarcity is likely to become much more serious in the future, compromising the achievement of the United Nations Sustainable Development Goals (SDGs).

That being said, Paulo Pereira's book is timely and extremely helpful. It discusses key issues related to climate change and urban soil and water degradation and management. Special attention is devoted to flooding, including groundwater quality and quantity, land and soil protection and policy, and ecosystem services. This book contains eight chapters written by experts in this field, some of them already coauthors of chapters in Volume 4 of this series.

In short, I would recommend both volumes to readers, to better learn and understand the general soil degradation problems discussed in Volume 4, followed by Volume 8, covering the future urban water and soil-related challenges. Both books will be very useful for the present and next generations who are planning to move from rural areas to megacities. The main problems to be solved were identified: lack of water and a climate change scenario. The final aim of Paulo's book, as he mentioned in his preface, is to move forward and improve the sustainability and resilience of the urban environment in the years to come. Therefore there is an urgent need to further improve water-use efficiency, control urbanization in water-scarce areas, mitigate water availability decline due to climate change, and undertake integrated sustainability analyses of potential solutions to address urban water scarcity and promote sustainable development.

Finally, I would like to thank all the recognized authors in this field for preparing their timely, practical, and easily readable chapters. This volume is highly recommended for soil scientists, hydrologists, urban planners, ecosystem service and climate change experts, and for newcomers and graduate students who want to learn more about soil degradation and hydrological issues in urban stressed environments. I am sure that all of them will enjoy this insightful and enjoyable volume!

Damia Barcelo
IDAEA-CSIC and ICRA-CERCA, Barcelona and Girona,
October 3, 2022

Volume editor's preface

More than half of the global population currently lives in urban areas, and this is expected to reach 68% by 2050[1]. The current migration to urban areas, observed mainly in developing countries, is dramatically increasing urban sprawl, with negative impacts on ecosystems and their services.[2] Most of the urban expansion is made at the expense of croplands, grasslands, and forests, significantly reducing the supply of ecosystem services such as flood and erosion regulation, water purification, pollination, and food production.[3,4]

One of the most visible impacts of urbanization is soil sealing. Sealing is one of the most important forms of land degradation[5] and substantially hampers soil functions.[6] Urbanization increases soil compaction[7] and the accumulation of different contaminants in soil, such as metals,[8] polycyclic aromatic hydrocarbons,[9] and microplastics.[10] Also, those managing urban green areas often use pesticides and herbicides.[11] The expansion of impervious surfaces affects the hydrology of urban areas. These surfaces drastically decrease water infiltration and increase the overland flow.[7] This is the most important cause of city flooding.[12] Floods are one of the most dangerous hazards[13] and are expected to increase with urban expansion and climate change.[14,15] The recent floods that occurred in Pakistan are one example of the dramatic impacts of this phenomenon, which affected approximately 33 million people[a]. The increase in flood frequency and severity is expected to enhance pollutants transported to surface and groundwater bodies.[16] Also, the increasing frequency and intensity of drought/heat wave events in many global regions are increasing water scarcity and insecurity in urban areas, where consumption is increasing quickly.[17]

Overall, urban soil and water degradation have increased due to urbanization and climate change. This trend is expected to impact human health and make cities less livable.[18,19] This is mainly observed in developing countries where the flux of people is high,[20] the expansion of slums is a severe problem,[21] the degradation of green areas is accentuated,[22] legislation is absent or is implemented weakly,[23] and climate change is expected to have a major impact.[24]

It is urgent to halt urban degradation, restore the ecosystems, and make the cities more livable.[25] Different global initiatives push the agenda to make urban areas more sustainable, such as the sustainable development goals

[a] https://public.wmo.int/en/media/news/climate-change-likely-increased-intense-rainfall-pakistan-study.

(Goal 11: Sustainable Cities and Communities; *Make cities and human settlements inclusive, safe, resilient and sustainable*),[26] the United Nations Decade on Ecosystem Restoration (2021–2030),[27] or the New Urban Agenda.[28] At the regional level, the new European Union Biodiversity Strategy 2030 focuses on ecosystems restoration, especially in greening urban and periurban areas.[29] Restoration in urban areas is essential to increase resilience to extreme events and increase the ecosystem services (e.g., water, air, and climate regulation; food production; recreation) supplied by these areas.[30] This will increase urban dwellers' wellbeing and quality of life.[31] When adapted to local conditions, green areas can be an excellent strategy for urban planning.[32] It is also essential that future developments consider the use of nature-based solutions[33] or the construction of houses with better energy efficiency that reduces consumption and carbon footprint.[34] These solutions are also essential to reduce the impacts of urban areas on climate change. It is imperative to include these principles in urban plans and legislation to increase urban areas' livability, resilience, and capacity to mitigate climate change.

We have to move forward in improving sustainability and resilience in urban environments. It is key to halt and reverse soil and water degradation in urban areas that is triggered by human activities and climate change. The hope that climate change is something that will come in the future is gone. Climate change is happening. If we do not adapt to this new reality and understand that nature, the natural capital, is the major asset we have to tackle climate change, our future will not be so prosperous and our health will be in danger. *The clock is ticking, and there is no way back.* With this book, we aim to contribute to the body of evidence on urban degradation and its processes.

Acknowledgements

The authors acknowledge the support and outstanding professionalism of Prof. Damia Barcelo, acquisitions editor Jason Mitchell, and developmental editor Jhon Michael Peñano in the development of this book volume. We greatly appreciate their efforts. We also want to acknowledge the authors who have contributed with chapters and who helped us make this book a reality.

Paulo Pereira
Environmental Management Laboratory,
Mykolas Romeris University, Vilnius, Lithuania

Carla Sofia Santos Ferreira
Department of Physical Geography and Bolin Centre for Climate Research,
Stockholm University, Stockholm, Sweden

Research Centre for Natural Resources, Environment and Society (CERNAS),
Coimbra Agrarian Technical School, Polytechnic Institute of Coimbra,
Coimbra, Portugal

References

1. United Nations, Department of Economic and Social Affairs, Population Division. *World urbanization prospects: the 2018 revision (ST/ESA/SER.A/420)*. New York: United Nations; 2019.
2. Pereira P. Ecosystem services in a changing environment. *Sci Total Environ* 2020;**702**:135008. https://doi.org/10.1016/j.scitotenv.2019.135008.
3. Leitão IA, Ferreira CSS, Ferreira AJD. Assessing long-term changes in potential ecosystem services of a peri-urbanizing Mediterranean catchment. *Sci Total Environ* 2019;**660**:993–1003. https://doi.org/10.1016/j.scitotenv.2019.01.088.
4. Pereira P, Inacio M, Kalinauskas M, Bogdzevič K, Bogunovic I, Zhao W. Land-use changes and ecosystem services. In: Pereira P, Gomes E, Rocha J, editors. *Mapping and forecast land use/cover changes. The present and future of planning*. Elsevier; 2022. p. 1–27. https://doi.org/10.1016/B978-0-323-90947-1.00007-7.
5. Xiao R, Tian Y, Xu G. Spatial gradient of urban green field influenced by soil sealing. *Sci Total Environ* 2020;**735**:139490. https://doi.org/10.1016/j.scitotenv.2020.139490.
6. Tobias S, Conen F, Duss A, Wenzel LM, Buser C, Alewell C. Soil sealing and unsealing: state of the art and examples. *Land Degrad Dev* 2018;**29**:2015–24. https://doi.org/10.1002/ldr.2919.
7. Ferreira CSS, Walsh RPD, Ferreira AJD. Degradation in urban areas. *Curr Opin Environ Sci Health* 2018;**5**:19–25. https://doi.org/10.1016/j.coesh.2018.04.001.
8. Gabarrón M, Faz A, Acosta JA. Use of multivariable and redundancy analysis to assess the behavior of metals and arsenic in urban soil and road dust affected by metallic mining as a base for risk assessment. *J Environ Manage* 2018;**206**:192–201. https://doi.org/10.1016/j.jenvman.2017.10.034.
9. Burant A, Selbig W, Furlong ET, Higgins CP. Trace organic contaminants in urban runoff: associations with urban land-use. *Environ Pollut* 2018;**242**:2068–77. https://doi.org/10.1016/j.envpol.2018.06.066.
10. Wang F, Wang Q, Adams CA, Sun Y, Zhang S. Effects of microplastics on soil properties: current knowledge and future perspectives. *J Hazard Mater* 2022;**424**:127531. https://doi.org/10.1016/j.jhazmat.2021.127531.
11. Ignatieva M, Hedblom M. An alternative urban green carpet. *Science* 2018;**362**:148–9. https://doi.org/10.1126/science.aau6974.
12. Sohn W, Kim JH, Li MH, Brown RD, Jaber FH. How does increasing impervious surfaces affect urban flooding in response to climate variability? *Ecol Indic* 2020;**118**:106774. https://doi.org/10.1016/j.ecolind.2020.106774.
13. WMO. *WMO atlas of mortality and economic losses from weather, climate and weather extremes (1970-2019)*. Geneva 2, Switzerland: World Meteorological Organization; 2021.
14. Kalantari Z, Ferreira CSS, Walsh RPD, Ferreira AJD, Destouni G. Urbanization development under climate change: hydrological responses in a peri-urban Mediterranean catchment. *Land Degrad Dev* 2017;**28**(7):2207–21. https://doi.org/10.1002/ldr.2747.
15. Winsemius H, Jongman B, Veldkamp T, Hallegatte S, Bangalore M, Ward P. Disaster risk, climate change, and poverty: assessing the global exposure of poor people to floods and droughts. *Environ Dev Econ* 2018;**23**:328–48. https://doi.org/10.1017/S1355770X17000444.
16. Mishra A, Alnahit A, Campbell B. Impact of land uses, drought, flood, wildfire, and cascading events on water quality and microbial communities: a review and analysis. *J Hydrol* 2021;**596**:125707. https://doi.org/10.1016/j.jhydrol.2020.125707.
17. Hoekstra AY, Buurman J, van Ginkel KCH. Urban water security: a review. *Environ Res Lett* 2018;**13**, 053002. https://doi.org/10.1088/1748-9326/aaba52.
18. Maurya SP, Singh PK, Ohri A, Singh R. Identification of indicators for sustainable urban water development planning. *Ecol Indic* 2020;**108**:105691. https://doi.org/10.1016/j.ecolind.2019.105691.

19. Wang P, Qiao W, Wang Y, Cao S, Zhang Y. Urban drought vulnerability assessment—A framework to integrate socio-economic, physical, and policy index in a vulnerability contribution analysis. *Sustain Cities Soc* 2020;**54**:102004. https://doi.org/10.1016/j.scs.2019.102004.
20. Sedova B, Kalkuhl M. Who are the climate migrants and where do they go? Evidence from rural India. *World Dev* 2020;**129**:104848. https://doi.org/10.1016/j.worlddev.2019.104848.
21. Patel A, Shah P, Beauregard B, B.E. Measuring multiple housing deprivations in urban India using Slum Severity Index. *Habitat Int* 2020;**101**:102190. https://doi.org/10.1016/j.habitatint.2020.102190.
22. Das M, Das A, Pereira P, Mandal A. Modeling spatio-temporal dynamics of ecosystem health (EH): a study on a rapidly urbanizing metropolitan area of lower Gangetic plain, India. *Ecol Indic* 2021;**125**:107584. https://doi.org/10.1016/j.ecolind.2021.107584.
23. Ren C, Yu CW, Cao SJ. Development of urban air environmental control policies and measures. *Indoor Built Environ* 2022. https://doi.org/10.1177/1420326X221120380.
24. Fawzy S, Osman AI, Doran J, Rooney DW. Strategies for mitigation of climate change: a review. *Environ Chem Lett* 2020;**18**:2069–94. https://doi.org/10.1007/s10311-020-01059-w.
25. Ferreira CSS, Kalantari Z, Pereira P. Liveable cities: current environmental challenges and paths for urban sustainability. *J Environ Manage* 2021;**221**:111458. https://doi.org/10.1016/j.jenvman.2020.111458.
26. United Nations. *Transforming our world. The 2030 agenda for sustainable development.* United Nations; 2015.
27. United Nations. *The United Nations decade on ecosystem restoration.* United Nations; 2021.
28. United Nations. *The new urban agenda.* United Nations; 2017.
29. European Commission. *Communication from the commission to the European parliament, the council, the European economic and social committee and the committee of the regions. EU Biodiversity Strategy for 2030. Bringing Nature to Our lives.* European Commission; 2020.
30. Pereira P, Inacio M, Karnauskaite D, Bogdzevič K, Gomes E, Kalinauskas M, et al. Nature based solutions impacts on urban environment chemistry: air, soil and water. In: Ferreira CSS, Kalantari Z, Hartmann T, Pereira P, editors. *Nature-based solutions for flood mitigation: environmental and socio-economic aspects.* Springer Publishing House; 2022. p. 79–137. https://doi.org/10.1007/698_2021_760.
31. Pinto LV, Inacio M, Ferreira CSS, Ferreira AD, Pereira P. Ecosystem services and wellbeing dimensions related to urban green spaces—A systematic review. *Sustain Cities Soc* 2022;**85**:104072. https://doi.org/10.1016/j.scs.2022.104072.
32. Pereira P, Baro F. Greening the city: thriving for biodiversity and sustainability. *Sci Total Environ* 2022;**817**:153032. https://doi.org/10.1016/j.scitotenv.2022.153032.
33. Raymond CM, Frantzeskaki N, Kabisch N, Berry P, Breil M, Nita MR, et al. A framework for assessing and implementing the co-benefits of nature-based solutions in urban areas. *Environ Sci Policy* 2017;**77**:15–24. https://doi.org/10.1016/j.envsci.2017.07.008.
34. Huo T, Li X, Cai W, Zuo J, Jia F, Wei H. Exploring the impact of urbanization on urban building carbon emissions in China: evidence from a provincial panel data model. *Sustain Cities Soc* 2020;**56**:102068. https://doi.org/10.1016/j.scs.2020.102068.

CHAPTER ONE

Climate change and property rights: Challenges for land policy

B. Ayça Ataç* and Thomas Hartmann
TU Dortmund University, Dortmund, Germany
*Corresponding author: e-mail address: ayca.atac@tu-dortmund.de

Contents

Abstract

The increasing number of environmental disasters such as floods, landslides, sea level rise, and storms have an impact on how land can be used and affect property rights. The effects are predominantly negative and often result in damage of property. The academic debate on land policy and its attempt to deal with such impacts are mainly responding to the economic aspect of robust property rights in land. Compensation mechanisms, tradable development rights, and relocation of residents are some of such land policy responses. However, there is a mismatch between the economic approach of land policy and plural notions of property rights. Such plurality arises from the property not being merely an economic asset, but also serving as social or cultural assets, symbolic attachment to a specific pieces of land, family legacy, or valuable place for other non-economic reasons.

This contribution explores the challenges of land policy in the face of climate change. The reasons behind the challenges of property rights to adjust them to current developments are revealed by discussing its nature. Then conclusions are drawn about how climate change reveals the plural notions of property in land through the perception of damage. As a result, this contribution shows the mismatch between the plural notions in landowners' perspectives and land policy's predominant economic approach. The main contribution is to demonstrate how various landowners' perspectives matter to land policy when the effects of climate change on urban areas are inevitable.

Keywords: Climate change, Land policy, Property rights, Landowners' perspective, Damage of property

Advances in Chemical Pollution, Environmental Management and Protection, Volume 8
ISSN 2468-9289
https://doi.org/10.1016/bs.apmp.2022.10.013

1. Introduction

Climate change impacts land use by affecting environmental conditions, particularly in urban areas. Consequences are increasing floods, landslides, droughts, storms, sea level rise, and other natural disasters.[1] These undermine the very foundation of land use and, as a result, property rights.[2]

While effects of climate change on property rights can sometimes be positive, such as discovering new resources, better crop growth in specific regions, or improving trade routes, in the majority of instances, they are negative. Changing environmental conditions often disadvantage landowners or deprive them of livelihood altogether.[2] According to the International Governmental Panel on Climate Change's special report,[3] extreme weather events are likely to become more frequent and catastrophic as a result of climate change. Flooding of coastal communities because of global sea level rise, pollution of underground and surface water supplies, or damage of built-infrastructure (e.g., houses, roads, railways, and electrical lines) due to increased hurricane and tropical storm intensity are some examples of how the effects of climate change manifest themselves abruptly.[4]

Climate change's various impacts on existing land use can be "long-term and gradual" or "temporal and recurring." Displacement or economic downturn are some long-term outcomes. For instance, due to climate change in India's various agro-climatic regions, property re-evaluation has become inevitable.[5] Changing precipitation and drought caused a major productivity decrease in agriculture, which is the most important sector in the country. Farmers who were unable to live with their basic means of subsistence began to migrate to urban areas. This triggered significant societal changes. In addition to migration from rural to urban areas, decrease in agricultural land value, loss of soil fertility and abandoning traditional agricultural production are other economic consequences of climate change. Another example of climate change's increasing impact which triggers changes on land use are floods. They are among the most common temporal and recurring results of environmental change. Floods are one of the costliest climate-related disasters,[6] as well as a risk to urban inhabitants. Many areas across Europe are expected to have an increase in precipitation as a result of climate change.[3] In the last three decades, it has impacted millions of people and resulted in massive economic losses.[7] This has consequences for the way, land can be (re-)used after disasters or in order to mitigate the impact of climate-induced risks. So, property rights in land can be affected by climate change.[2]

However, property rights do not necessarily disappear. Impacts can also entail restrictions of land use for example in the floodplains,[8] or constraints on the economic viability of certain land uses. Governmental authorities can impose legal sanctions to limit property rights or the use of property. Some government strategies, such as building restrictions in floodplains, include enforcing mandatory construction regulations and prohibiting urban development in specific areas.[9] So, the impact of climate change on property in land results not only directly from changing environmental conditions, but also political decisions on different levels of government can affect property in land—for example by restricting certain land uses in order to mitigate or adapt to climate change. "The actions of government agencies in restricting land rights for (prospective) climate change risks are at least as important as the actual disappearance of rights due to permanent flooding or storm events."[2] The interventions of public authorities can vary in the degree to which the influence property[10]—while some measures merely regulate land uses using public policy without impacting property rights (such as incentives or subsidies for certain measures), other interventions impact or even redefine the scope and the content of use or disposal rights (such as regulations or land-use planning). Resettlements and evacuations are rather strong governmental interventions, as they restrict redistribute property rights in land. Such strategies attempt to permanently avoid possible damage.[11] Relocating individuals from high-risk areas to other locations is a long-term effective measure. However, lack of political will, lack of regulations and financial compensation, uncertainty about the process and outcome of the planned relocation, and psychological factors (transfer), such as emotional reactivity, grief, and exhaustion, are all common obstacles to this measure.[12] Therefore, these strategies are not widely spread.

Thus, as mentioned above climate change can affect property in land in a number of ways, including biophysical, economic, and legal. These effects have negative consequences for cities at short and long term. This contribution considers the effects of climate change on property rights from landowners' perspectives. This contribution discusses how different landowners perceive the impact of climate change and its consequences differently. It will be elaborated how different meanings can be assigned to property in land. The chapter further discusses respective land policy interventions and how it can fall short of providing adequate solutions to the multiple perceptions of property in land by landowners as land policy often considers private property as an economic asset. It is concluded how, considering multiple landowner perspectives on property in land, there can emerge a mismatch between land policy's mainly economic approach and multiple landowner perspectives.

2. The nature of property rights

Property rights are robust in their nature. On the one hand, this robustness provides a basis for landowners' welfare security such getting mortgages or secure investments in land. On the other hand, it conflicts with public interest when flexible adjustments and amendments of land use are required to address environmental problems. Property rights are robust because "often it takes a long time to change them. One reason is that they represent deeply rooted moral beliefs. Many of the changes are fairly marginal and have the form of forbidding what previously was a presumptive right."[13] As a legal right established by a combination of private and public legislation, property rights are difficult to change. When planning based on law and property rights tries to be flexible, it runs the risk of handling people's property rights haphazardly.[14] Therefore, when there is a need to make changes in property rights, the legal and moral limitations have to be considered.

Understanding the roots of property rights shows how it has been regarded an absolute right in the past but requires changes now. Property rights were discussed in the 18th century by political philosophers such as John Locke, and Jean-Jacque Rousseau, or political economist Adam Smith. These philosophers have discussed of how democratic governance structures and market economies can be established and implemented. Their thoughts were focused on creating a strong and enforced set of privately owned property.[15] In conclusion, property rights have developed as a robust human rights.

Despite differences in its specific implementations in different constitutions, property rights are legally significant and constitutions all around the globe protect them rather well in liberal democracies. For instance, the Fifth Amendment of the US Constitution protects the right to private property, stating: "...nor shall private property be taken for public use, without just compensation." The constitution also reveals the connection between liberty and property, inspired by Locke: "No person shall ... be deprived of life, liberty, or property, without due process of law" (The Constitution of the United States, Amendment 5). Property rights are also mentioned in the first protocol of the European Convention on Human Rights as well. It states that "Every natural or legal person is entitled to the peaceful enjoyment of his possessions. No one shall be deprived of his possessions except in the public interest and subject to the conditions provided for by law and by the general principles of international law" (Council of Europe, 1950,[16] art. 10.1). According to this statement, property rights, which are widely regarded as one of the most fundamental human rights, can be directly linked to individual liberty. Besides, as emphasized by Jacobs,[15] 18th century philosophers inspired legal documents that link liberty and property rights. This close connection can be considered as the most primary indicator of property rights' absolute structure.

Despite its robustness, right to own a property protects the individual against arbitrary or arrogant actions of politicians, state officials and other holders of power.[13] "This is the most important guarantee of freedom, not only for those who own property, but scarcely less for those who do not."[17] In fact, "in the history of human rights, the concepts of 'freedom' and 'property' have been interrelated."[18] These interpretations reveals the strong relationship between human rights and property rights.

According to John Locke, individual rights, especially the right to own and manage property, are protected by freely constructed governments. Locke, together with Rousseau, had a major influence on how people conceived about the content of the American Revolution. Human rights ideas were evolving, and they were closely intertwined to ideas about private property. James Madison (the fourth American President) and John Adams (the second American President) indicated the importance of property rights and its protection by the government. According to Adams, "property must be secured or liberty cannot exist."[19] The strong connection between human rights and property rights is seen as a guarantee of individuals' freedom.

Of course, the freedom that comes with owning property is not without limits. Talking about the limits of property, Karl Marx stated that "The right of property, work, ... the right to enjoy one's fortunes ... without regard for other men and independently of society ... It leads every man to see in other men ... the limitation of his own liberty" (Marx, 1840, as cited in Ref. 15). Ownership is not only a right but also a responsibility. Article 14 paragraph 1 of the Basic Law of the Federal Republic of Germany (Grundgesetz, GG) guarantees private property. Paragraph 1 states that "Property and the right of inheritance shall be guaranteed. Their content and limits shall be defined by the laws" (article 14 I GG). In the second paragraph of the same article, it is stated that "Property entails obligations. Its use shall also serve the public good" (article 14 II GG). Therefore, property rights give freedom to landowners, but they also show the limits of their own freedom. This is one of the indications why property rights are not absolute.

All these statements made by various scholars such as Locke, Rousseau, Smith, and Marx allow us to justificate the robust structure and necessity of property rights by using their relationship with human rights. Even for the common good, making some changes in this social function that is directly related to human rights may cause chaos in society. However, when interpreting these statements today, the environmental problems of today's world, in addition to the rate of urbanization and industrialization, have to be taken into account. The robustness of this social function (property rights) frequently appears as a barrier to combating the effects of climate change. This is because, as land became more urbanized and industrialized, the government began to restrict people's rights to use and develop their land. Conflicts arose over public rights on private property.[20] In this case, robust set of private property rights is faced with diverse set of natural resources and issues.

Property rights, is traditionally considered absolute, "the widest right granting the owner the power to do with his property whatever he may wish."[21] Also classical economist David Ricardo states that power over land is indestructible. According to him, "rent is that portion of the produce of the earth, which is paid to the landlord for the use of the original and indestructible powers of the soil."[22] Nevertheless, occasions are emerging today where it may be necessary or appropriate to restrict property rights to cope with results of changing environmental conditions. In this case, "indestructible land" mentioned by Ricardo[22] is not that indestructible anymore. Making changes in land use to combat the effects of climate change or to take precautions against the risks that may arise from climate change can lead to restrictions on property rights.

Reflecting on property rights, Harvey Jacobs[15] states that Garrett Hardin and Hernando De Soto support more private property rights in the age of climate change and urbanization. More property rights is the solution to environmental and urban problems. They advocate privatization of natural resources. The basis for this is that resource owners wish to manage their property for long-term sustainability, which includes avoiding liability for resource misuse or abuse. The arguments of Locke and Rousseau are acknowledged by Hardin and Soto. They claim that a properly defined and widely distributed private property can achieve the best results in terms of sustainability.

Thus, as highlighted above, property rights are one of the basic human rights and they are directly linked to the individual's freedom and liberty. Since 18th century, they are discussed by philosophers and politicians. Moreover, they are mentioned in all sort of legal documents and included in the constitutions in liberal democracies. Making changes on this right that is strongly protected by laws is difficult and often takes long time. Due to these reasons, the robust and inflexible nature of property rights are discussed this section. Besides, its well-protected structure and how this structure provides a welfare security for landowners are explained. Nevertheless, today's environmental conditions are different than 18th century. Climate change and its increasing effects on cities oftentimes require changes on land uses, therefore on private property. Considering the nature of property rights, its legal and moral limitations have to take into account when applying changes on them.

3. The damage of property rights

The effects of climate change on property rights and their robust nature[14] have been explored so far. In this section of the contribution, the impact of climate change on property rights will be evaluated from perspective of damage. The impact of climate change on urban areas necessitates the participation of many different actors in the management process, such as public and private institutions, small-scale businesses, landowners, tenants, and so forth. All of these actors interact with one another through a complex and entangled set of constitutional, statutory, and common law legal doctrines, as well as other public institutional structures and private market forces.[23] This interaction mentioned by Norton and Meadow between different actors emerges as a tension when it comes to find the balance between the robustness of property rights and adopting land use changes. On the one hand, in order to combat the effects of climate change and mitigate negative consequences, land use changes and thus interference with property rights are necessary. On the other hand, because of its constitutionally protected structure, removing or changing property rights after they have been assigned is difficult. Landowners are in the center of this discussion. They own the property and they are affected from both environmental changes and mitigation measures on their land.[2] Therefore, they have a strong impact in the risk management process. Landowners may either suffer from the effects of climate change or have their rights on their property affected due to adaptation measures. Therefore, landowners experience damage to their land and property rights.

Damage can be considered a reduction in the value or usefulness of assets. Such a conception of damage implies that value or use needs to be attached to potentially damaged objects. Often, this impact is evaluated mainly as economic and physical. However, there are multiple interpretations of the damage, and its definition depends on how it is interpreted. Effects on property (rights) vary according to the value different landowners ascribe to their property. The economic approach is the most common one. It is the consideration of any damage to property caused by the effects of climate change through monetary instruments. Referring to real estate appraisal, most instruments of land policy, such as expropriation, preemption, and contractual agreements, treat land mainly as an economic asset. This approach to damage alludes to direct and indirect financial losses by damage to property assets, basic materials, and goods. Risk management and compensation methods across the globe, for example in the European Union (EU), its Member States, the United States, and England reflect this prevalent perception. A study which compares post-flood policies and compensation regimes in the EU and the US focuses on monetary instruments such as flood insurance, government payments, and resilient reinstatement grants.[24] In addition to the economic aspects, physical aspects like building regulations are considered by the countries during the risk management process. However, Suykens' comparative study, proves that solely economic grants have been unsuccessful in the past.

Thaler and Fuchs[12] investigated the effects of financial payments in Austria on the improvement of individual landowners' and businesses' encouragement to prepare their properties for disaster risk. The results of the research revealed that financial resources do not provide a direction that will encourage individuals to avoid future losses or change their risk management behavior. In China, residents are only partially reimbursed for direct losses such as housing and agricultural crop losses. Damages other than economic, such as psychological trauma or a failure to maintain the property, are not taken into account.[25] In their research on the distributional effects of flood risk management, Dai et al.[25] outlined an unclear, insufficient, and ad hoc compensation mechanism for flood retention areas in

China. With a similar approach, underestimating different aspects such as social, cultural, and ecological in the economic approach might result in insufficient compensatory mechanisms for other countries as well. Monetary instruments are not adequate by itself to avoid damage on properties. Another example demonstrating the shortcoming of solely economic approach is coastal communities in Louisiana.[26] They are facing gradual disasters namely erosion and sea level rise. Sand[26] explores how economic decision-making methods, such as cost-benefit analysis, used in solving problems in this area have led to the loss of Louisiana's unique cultural aspects. Despite repeated disasters and property losses, local owners avoid relocation. This is due to the strong connection people have with culture and society and the fact that the place they live becomes a part of their identity.

Apart from the above-mentioned merely economic approaches, property rights are a social notion that is closely related to spatial attachment. This attachment of individuals is an important factor in their decision-making on their property. Results of a study in India revealed that place attachment was significantly correlated with flood preparedness.[27] The example of Orissa, which is economically the second lowest state in India, suffers from severe floods, loss of people, livestock and human settlements. Residents in Orissa rely on their ancestral possessions, especially their land, home, and other physical property. Their livelihoods are location-dependent, implying that there may be economic place attachment, which might impact disaster preparedness.[27] Based on this example from Mishra et al.,[27] when landowner sees his or her property as "home," then he or she is dependent on the neighborhood. It is the territory that they want to protect. This attachment is not just to an economic asset, but also to the memories that have been accumulated in the house, to the neighborhood, and to the social environment. Therefore, in addition to the property's economic viability, its social structure should also be considered when combating the effects of environmental change.

In addition to the economic and social interpretation of property rights, customary land rights and indigenous notions of property should be thoroughly considered to avoid disruptions and injustice. For instance, research conducted by Otsuki and Jasaw[28] in Ghana, West Africa, demonstrates that customary tenure arrangements rooted in Africa's social and historical context have resisted land reform efforts involving formalization and privatization. It is difficult to change these customary land rights, which are deeply rooted in the social structure and different from western planning theory and laws. Therefore, property rights must be viewed as a form of institution that is responsive to both customary arrangements and individual and social adaptation requirements.

Clearly, economic value is one of the most important features of property rights. They can be bought and sold. Moreover, every intervention on properties, including disaster risk mitigation, insurance, or renovation, has a monetary value. However, this is not the only aspect of property (rights). This aspect changes according to who is assessing or measuring any damage that has occurred or may occur to the property. In other words, the value given to the property and the decision of who the winner is and who the loser is depend on the measurement criteria of the damage. Needham[13] explains this perception of damage with the example of the fish-and-chips shop. "If we are living in a quiet street and a fish-and-chips shop wants to locate next door, we think we have the right to continue to live there undisturbed by smell and eaters on the street. We expect the courts to protect our rights." Only if imposed by a government agency, such as a change in land-use plan do we expect a reduction in property rights. Property rights cannot be expected to be abolished for the sole purpose of welfare maximization. Furthermore, Coase[29] details the damage aspect with the fish example and asks the question: "If we assume that the harmful effect of the pollution is that it kills the fish, the question to be decided is, Is the value of the fish lost greater or less than the value of the product which the contamination of the stream makes possible?" Therefore, money costs and environmental contamination are two dimensions of this damage described by Coase.[29]

As mentioned from the very beginning of this section, damage to property is mainly regarded by state authorities as an economic asset. Monetary instruments are typically utilized for intervention and compensation.[30] However, when looking at property damage from a landowner's perspective, loss of property or forced relocation as a result of property damage is not just an economic loss. This harm occurs in the loss of social connection for the landowner who is socially and emotionally engaged with their home. When the perception of damage is examined in indigenous notions of property rights, it becomes clear that even minor changes in property rights have an impact on the foundations of customary land rights.[2]

Thus, mainly an economic interpretation of property rights is insufficient to assess and respond to the impact of climate change on property rights. The way property rights are affected by environmental changes and how to address related challenges depend on the location-specific social construction of property rights. Changing environmental conditions reveal different layers of property rights beyond its monetary value, namely social, traditional, and customary rights. They show the diverse and plural notions of property rights. In this part of the contribution, perception of damage was used to reveal the differences in the value landowners ascribe to their property. Climate change fundamentally affects property rights. Its effects bring the damage on private property and the plurality of landowners' perspectives to light. Without neglecting the plural notions of property in land, monetary compensation alone remains insufficient. Plurality requires more responsive, diverse, and comprehensive land policy approaches. This has implications that current policy interventions do not consider property rights as a multifaceted social function.

4. Challenges and responses of land policy

The impact of climate change on land use and eventually on property rights challenges the idea of property rights in land being socially constructed as robust and almost indestructible.[2] This reveals inherent and entrenched believes and conceptions of landowners' perspective on property in land and how these landowners designate different values to their properties. While some consider their property as a source of investment, some consider it as a home or others as status symbol.[31] They are thus attached to the property in different ways. Land policy decisions that overlook such different perspectives on property in land fail to respond to the needs of landowners.[32] Therefore, this section will cover challenges led by multiple landowners' perspectives on land policy as well as theoretical approaches to address these challenges.

Although landowners interpret damage to their properties in various ways, state authorities focus primarily on the economic perspective when deploying land policies to respond to this damage. The economic approach has undeniable importance as a compensation mechanism.[33] Nevertheless, this approach addresses only one of the several landowners' perspectives. Other perspectives, such as social, spatial, or emotional, matter to landowners as well. For example, a landowner's property may have an economic value. When faced with changes to his or her property, the landowner wishes to preserve the economic interest. For the same landowner, the very same property can also serve as a family gathering place. In the event of a relocation or evacuation, the landowner is concerned about losing both the property's economic value as well as his or her home, which has many personal attachments.[12] Therefore, economic compensation alone is not sufficient to subsidize the landowners' perceived damage. Damage, in other words, is not merely an economic concept.[34]

Furthermore, there is overlap and interaction between landowners' multiple perspectives. These perspectives accumulate on a piece of land as a result of damage to property. Based on an exploratory study in Turkey, landowners of properties that were subject to flooding in 2021 have been interviewed. The interviews revealed a variety of meanings that landowners ascribe to their properties. One landowner, for example, describes his property as a "comfort zone" because it is close to his work and his children's school. The same landowner also emphasizes the importance of state-provided financial compensation, moral support, and citizen collaboration. This landowner is most concerned about the impact on his property's strategic location and economic value. For him, moral compensation is just as important as financial support. As a result of the same flood event, another landowner who was born and raised in the same neighborhood stated that he lost his home, his family gathering place, his memories attached to the property, and his neighbors. He emphasized the reason for leaving his property behind as having heavy emotional trauma related to great moral loss. This landowner's decision to leave his property as a result of the trauma reflects his perception of the non-physical damage he and his family experienced. When discussing property damage, this landowner makes no reference to economic loss. These examples demonstrate how different landowners perceive property damage in different ways. Multiple perspectives on property damage emerge even by taking one landowner's perception. Therefore, landowners expect diverse land policy approaches that respond to these multiple perspectives.

The overlap and interaction between landowners' perspectives cause problems for land policy. The reason has to do with the aforementioned multiple perspectives of landowners on their properties. Land policy often fails to respond to all such perspectives. This can lead to less effective and less efficient land policy interventions, as landowners' responses may not answer to the rationale of the policy intervention (i.e., a monetary incentive is not able to trigger certain action, if the action is motivated not for monetary reasons, see Ref. 32).

Managing the tension between different actors is a difficult task for land policy because of the multiple perspectives. At the same time, landowners are the main actors in climate mitigation and adaptation.[35] When land policy attempts to focus on each landowner's perspective individually rather than addressing property damage as a whole, policy responses are unable to meet all expectations, as known from other land-intensive policy fields, such as energy transition.[36] An attempt to approach such complex systems, in which multiple actors and their relationships matter, is complexity theory.

Complexity theory contributes to the understanding of land policy issues that are uncertain and difficult to identify. Cairney[37] summarizes complexity theory's four key policy-making insights. These insights are explained together with the relevance of using complexity theory in this research. First of all, applying law-like behavior in a complex system is difficult due to the involvement of various forces. For example, a land policy response that was successful for landowner A may not have the same result for landowner B. The idea of complexity theory is that success is the result of a variety of factors, each of which is specific to the situation. Considering different variables matter for different landowners, it is difficult to apply the same intervention to another landowner. Second, self-organizing systems are difficult to control. It is impossible to predict how internal and external forces will act on the complex system. Therefore, when land policy decisions over landowners do not have the desired effect, this is not exactly unexpected. There is an inevitable degree of "error" in land policy. Landowners' perspectives can be influenced by external factors, regardless of the intention of land policy decisions on private properties. Third, complex systems are inherently unstable and frequently change. Land policy responses must consider this dynamism and not rely on a single policy strategy. Last but not least, actors in complex systems have their own perceptions and decide how to behave. On their properties, policy decisions can be made. However, unless landowners collaborate, land policy's power is insufficient to resolve issues involving their properties and, on some occasions, the common good.

Complexity theory suggests shifting in policy analysis from individual parts of a system to the system as a whole. A system needs to be examined as a whole, rather than as an assembly of distinct and separate elements because interaction of the elements has an impact on the whole. This statement helps us to focus on bringing multiple landowners' perspectives together. In line with this concept, because each actor provides a unique value, bringing the perspectives of multiple landowners together contributes to the decision-making process. To better define land policy problems as a whole, de Roo[38] suggests a spectrum that includes two poles, namely technical and communicative rationales. The localization of the problem in this spectrum depends on its certainty-uncertainty ratio. The transformations occurring along the spectrum, moving from technical to communicative rationale, provides insights into land policy responses when tackling climate change.

The technical extreme, the first theoretical pole in the spectrum, represents certainty. Technical interventions are needed here to reach the desired result. It has a certainty that is clear and well-defined. Land policy's solely economic responses to climate-related issues can be considered closer to this technical pole because of its rather well-defined structure. When other aspects do not have clear definitions, it is easier to grasp the economic one and design solutions for it. The communicative extreme, which is the other theoretical pole, involves varying actors and interaction between them. It causes this extreme to be uncertain. Plural perspectives of property in land can be associated with this uncertain pole. Due to the effects of climate change on private property, many landowners, their multiple expectations and interaction between these expectations get involved in land policy decision-making process. The degree of uncertainty in the decision-making process increases as the number of actors involved in the process expands. Within the communicative extreme, uncertainty is the reality of the decision-making process, and relations between actors are critical.

Complexity theory can be helpful in understanding how traditional policy making fails to address complex issues. It helps to categorize and understand the tension arise from multiple landowner perspectives. However, it is not very operational to grasp land policy when it comes to dealing with property rights. Land policy needs rigid planning including real rules. When climate change and its effects become tangible land policy get involved to respond to this complex phenomenon. In addition to social and emotional effects, land policy have considerable economic effects on individuals' lives. "Land-use planning can make a person poorer or richer, by affecting the value of the rights on her land."[13] On the one hand, such arrangements can directly interfere one's property rights. On the other hand, it indirectly effects their individual rights by influencing what others can do on their land. Land-use decisions require explicit definitions because they have an undeniable economic impact on individuals. Therefore, when dealing with the plurality and multiple perspectives in land, simply understanding its complexity remains insufficient. There is a need for a way of dealing with the pluralism.

The Cultural Theory helps to go beyond merely stating property as complex and to discuss possible land policy interventions without neglecting the plurality. The plurality can be better understood by correlating multiple perceptions with the Cultural Theory. This theory provides a model which reduces the number of possible rational expectations from the landowners without neglecting the plurality in their perceptions. It can be used as a facilitator in analyzing the diversity in stakeholder expectations during the risk management process and damage perception. Furthermore, the Cultural Theory helps to resolve the uncertainties and complexities encountered.[39] Therefore, discovering the plural notions of property in land becomes possible with this manageable number of rationalities.

Take the example of landowners A and B, both of whom live in the same neighborhood. Landowner A owns the property for over 20 years. He or she lives together with his family and raised their children in this apartment. Landowner B bought the apartment strategically with the sole purpose of investment. These two landowners put different values on their properties. Certainly for both of them, the economic value of the property is important. Do landowner A and landowner B have the same perception of damage when relocation is inevitable for them because of flooding? How do they prioritize the damage on their properties? By moving to another property, landowner A not only loses the social network within the neighborhood, but also his or her memories attached to the place. Economic value of the property is only one of the concerns. Landowner B is primarily concerned with his or her economic interest. The location or rapidly increasing land values in the area may have influenced this landowner's decision to purchase this specific property in the first place. On the one hand, landowner B expects land policy decisions to protect his or her economic interests as a result of relocation. Landowner A, on the other hand, cannot be compensated solely through an economic approach due to their multiple expectations. This landowner has a strong social and emotional attachment to his property. Therefore, location-specific land policy responses that incorporate these multiple perspectives are needed.

None of these perspectives can be proven right or wrong. All of them are empirically true. Plural notions of property in land do not always appear rationally. From the perspective of the one, the other one is irrational. The Cultural Theory takes up this irrationality and rationalize it under four perspectives, namely egalitarianism, individualism, hierarchism, and fatalism (Fig. 1). This theory is useful for understanding controversial differences.

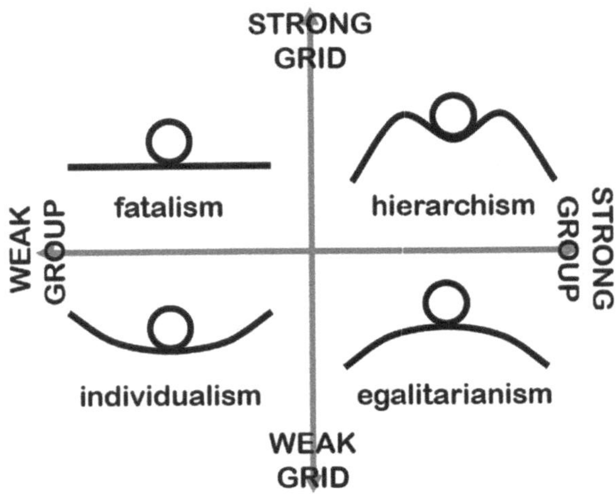

Fig. 1 The grid-and-group scheme of Cultural Theory. *From Hartmann T. Wicked problems and clumsy solutions: planning as expectation management.* Planning Theory *2012;**11**(3):242–256. https://doi.org/10.1177/1473095212440427.*

The loyalties and moral principles of individuals are based on the conflicts experienced in the solution of problems. Because of this, the disputes in question cannot be resolved easily.[39]

Individualism, the first of these rationalities, believes that the world seems to lie below in the valley. If a disturbance disrupts the equilibrium, it will always "roll" back to the valley's center. Every mistake made is seen as an opportunity to benefit. There is no compulsion to act together in a group.[40] As described by John Locke, property is an economic asset and a form of liberty for individualists.[15] Any damage to the property means a damage to their liberty. They follow market principles. According to this rationality, the most serious risk is inefficiency due to market failure. Since climate change effects their liberty, land policy needs to play its role to enhance economic welfare while also preserving their individual property rights. Individualists are open to experimentation.[40] Therefore, more experimental and responsive land policy is possible in this rationality. Here is an example of a landowner whose characteristics can be associated with the individualist rational. The landowner moved to the property for work reasons together with the family. This landowner does not have family or relatives living in the same neighborhood, but the property is close to the work place. The family considers it their comfort zone and they desire to protect it.

In addition to being in their comfort zone, the property's economic value is important to the landowner as well. When climate change's effect on landowners' property rights is inevitable, they expect to keep their comfort zone without losing its economic value.

The second rationale, *hierarchism*, assumes that the world is on top of a hill with a dip on top, resulting in a rather stable equilibrium. There is a certain degree of trial and error. This group prefers to set boundaries by establishing rules and regulations.[39] For them, property is more like a territory. It serves a certain function inside society. Any damage to the property means a damage to their social status. Hierarchists can be compensated with clear rule sets and procedures about damage when tackling the results of climate change. These rules should apply to everyone. Members of society give power to an institution, as in Thomas Hobbes' "Leviathan." As a result, strong group cohesion is essential.[40] All members of the group are on an equal level. Hierarchism requires a land policy that defines binding land-use decisions on expert knowledge and provides legal certainty. The characteristics of this rationality is similar to landowners' characteristics who strategically decide by seeking expert opinion when purchasing the property. In the event of any change in property rights due to the effects of climate change, these landowners expect land policy decisions to maintain their economic interests. Land use distribution and process management based on expert knowledge are other expected land policy interventions by these landowners. Before the law, landowner equality must be established.

Egalitarianism, the third of these rationalities, believes that the world is on an unstable hill. Even a minor disruption has the potential to irreversibly disrupt the balance. If an experiment fails, the equilibrium will be destroyed. So, there is no room for experiments. From this perspective, the outcome of an action, rather than the process, is more essential. Worry or moral responsibility are motivations to react; equality, democracy, and community are more important ideals than individual liberty or institutional integrity. According to this group, property is a form of social security. There is a moral motion and it entails obligations. Members of the group share a sense of community and work together to achieve their goals.[39] When dealing with damage, the community supports its members. Any damage to the property means a damage to their social stability and community. A landowner who holds the same property's ownership for a long time-period is an example to better explain the egalitarian perspective. Since the landowner was born and raised in the same neighborhood, he or she is spatially and emotionally connected to the social environment and neighbors as well as to the property. Social solidarity and cooperation are at the forefront for the landowner when changes in property rights are inevitable due to any damage caused by climate change. In addition to their property, these landowners are not willing to change the environment they are connected to. They expect land policy decisions to maintain the current status of their property within the framework of social justice.

Fatalism, the fourth and last rationality, is founded on the belief that we cannot predict how the world will react and that we have no control over events. For them, the inconceivability of the world arises from its complexity. Fatalists consider property just as nature and wildness. They believe on luck rather than justice. So, they do not believe that the world can be controlled.[40] For land policy, this group is challenging to handle because it represents the passive rationality of the four. Fatalists do not take land policy decisions into account because they do not believe in having an impact on the world. Climate change measurements to be taken on the land are not meaningful to them. Landowners that fall under this category are adamant about avoiding changing their properties. They refuse to comply voluntarily with any attempt to change their routine because they believe that any change will reduce their living standards.

Thus, various landowners perceive the effect, or the damage, on their property differently. Climate change is an example which reveals the plurality in landowners' perspectives when it comes to damage on their properties.[32] This section discussed multiple landowner perspectives and their interaction with each other. The interaction and overlap between different perspectives causes tension for land policy. When tackling climate-related issues on properties, solely economic land policy decisions are not sufficient to grasp multiple perspectives. Different landowners expect location-based decisions for their properties. Classification of the multiple perspectives can be managed with the help of complexity theory. However, when it comes to making changes on property rights, which is a legally well-protected human right, this theory lacks a clear set of rules. The Cultural Theory reduces the number of perceptions to four without neglecting its mutual contradictions.[39] This reduction is an advantage over complexity theory, because the acknowledgment of competing perceptions allows for a better comprehension of complexity than dichotomous models.

5. Discussion and concluding remarks

The relationship between climate change and property rights is substantial. This relationship arises from the fact that changing environmental conditions have a direct or indirect impact on an individual's property and property rights. The direct effects can occur dramatically when extreme events such as floods, sea level rise, or landslides physically damage the property. Indirect effects may arise from precautions taken to minimize or prevent potential damage to property caused by these weather events. An example of this effect is the relocation of property owners in a region declared a high-risk area after a disaster. In this scenario, the owner's title to the existing land is transferred to the new location determined by the public authorities without their choice. In this case, a change in property rights becomes inevitable when tackling climate-related issues.

However, as mentioned above, property rights are robust in nature. This robustness makes it difficult to deal with the problems arising from climate change. The reason for their robustness is that they are established as a combination of private and public legislations. For these reasons, they are difficult to change. Moreover, in many cases the desired changes take a long time. Property rights are well protected human rights in this respect. Although this structure provides the welfare security of the landowners, it conflicts with public interest when facing environmental issues. Landowners are concerned about their private interests and expect that decisions made throughout the planning process will be consistent with them. Planners, on the other hand, needs to develop land policy decisions that will be in the public interest as well as pursuing the interests of each landowner.

Oftentimes, considering the perception of damage, the impact of climate change on property rights is evaluated mainly as economic and physical. However, economic approaches alone are insufficient to explain land conflicts. The definition of damage depends on how it is interpreted. Its effect on property rights differs according to landowners' valuation criteria for their properties. Property can serve as a social or cultural value, a symbolic attachment to a specific piece of land, a family legacy, or a valuable place for other non-economic reasons. These are some of the plural perspectives of property in land that become more evident as a result of climate change. When dealing with the effects of climate change on private property, these plural notions require diverse land policy responses. Economic responses alone remain insufficient to address the plurality.

Current land policy responses such as compensation mechanisms, tradable development rights, and relocation of residents are mostly fiscal in nature, and they face challenges when tackling property-related issues. Therefore, there is a mismatch between the land policy responses and landowners' expectations. On the one hand, one of the main aims of planning processes is to regulate land-use that concerns private property and its economic value. Landowners, on the other hand, expect decision-makers to protect their private interests and welfare security. When the planning process overlooks the plural notions of property in land, it does not respond to the different expectations of different landowners.

This contribution argues that plurality in landowners' perspectives should be explored in depth in order to develop more efficient land policy responses and instruments. It is founded on the Cultural Theory and its four rationalities namely individualism, hierarchism, egalitarianism, and fatalism for comprehending various perspectives and complexities. In this research, various possible landowner perceptions are reduced into four categories by correlating distinct rationalities with their behavior. As a result, it is argued that the plural notions of property in land can be addressed by four alternative land policy responses. This contribution discusses different dimensions of the landowners' perception through the analysis of the notion of plurality.

The main idea of this contribution is to reveal how plural notions matter to land policy when the effects of climate change on private property are inevitable. To do so, it approaches the relationship between climate change and property rights from the landowners' perspective. The first step in addressing potential challenges is to consider the robust nature of property rights and the legal restrictions on them. Following that, the plural notions of property in land as a result of climate change are discussed. Damage to property is used as an illustration to reveal the plurality of landowners' perspectives. Multiple landowner perspectives and how to associate them with the Cultural Theory's four rationalities are discussed as a result.

References

1. IPCC. *Summary for policymakers (global warming of 1.5°C. An IPCC special report on the impacts of global warming of 1.5°C above pre-industrial levels and related global greenhouse gas emission pathways, in the context of strengthening the global response to the threat of climate change, sustainable development, and efforts to eradicate poverty)*; 2018.
2. van Straalen F. In: Hartmann T, Sheehan J, editors. *Property rights and climate change: land use under changing environmental conditions*. Routledge; 2018.

3. IPCC. *Climate change 2022: impacts, adaptation and vulnerability [IPCC WGII sixth assessment report]*; 2022.
4. Doremus H. Climate change and the evolution of property rights. *Univ Calif Irvine Law Rev* 2012;**1**:34.
5. Pellissery S, Praveena S. Climate change-induced property re-evaluation in agrarian contexts. In: van Straalen F, Hartmann T, Sheehan J, editors. *Property rights and climate change: land use under changing environmental conditions*. Routledge; 2018.
6. Hartmann T, van Doorn-Hoekveld W, van Rijswick M, Spit T. Editorial. *Water Int* 2019;**44**(5):489–95. https://doi.org/10.1080/02508060.2019.1671464.
7. Directorate General for Climate Action. *Consequences of climate change*; 2022. https://ec.europa.eu/clima/climate-change/consequences-climate-change_en.
8. Tarlock D, Albrecht J. Potential constitutional constraints on the regulation of flood plain development: Three case studies: potential constitutional constraints. *J Flood Risk Manage* 2016;**11**(1):48–55. https://doi.org/10.1111/jfr3.12274.
9. Hartmann T, Albrecht J. From flood protection to flood risk management: condition-based and performance-based regulations in German water law. *J Environ Law* 2014;**26**(2):243–68. https://doi.org/10.1093/jel/equ015.
10. Hengstermann A, Hartmann T. Instruments of land policy. Four types of intervention. In: Gerber J-D, Hartmann T, Hengstermann A, editors. *Instruments of land policy. Dealing with scarcity of land*. Routledge; 2018.
11. de Sherbinin A, Castro M, Gemenne F, Cernea MM, Adamo S, Fearnside PM, et al. Preparing for resettlement associated with climate change. *Science* 2011;**334**(6055):456–7. https://doi.org/10.1126/science.1208821.
12. Thaler T, Fuchs S. Financial recovery schemes in Austria: how planned relocation is used as an answer to future flood events. *Environ Hazards* 2020;**19**(3):268–84. https://doi.org/10.1080/17477891.2019.1665982.
13. Needham B. *Planning law and economics: an investigation of the rules we make for using land*. Routledge; 2006.
14. Hartmann T, Needham B, editors. *Planning by law and property rights reconsidered*. Ashgate; 2012.
15. Jacobs HM. Eighteenth-century property rights for twenty-first-century environmental conditions. In: van Straalen F, Hartmann T, Sheehan J, editors. *Property rights and climate change: land use under changing environmental conditions*. Routledge; 2018.
16. Council of Europe. *Convention for the protection of human rights and fundamental freedoms*. Council of Europe Treaty Series 005; 1950.
17. von Hayek FA. *The road to serfdom (Repr)*. Routledge; 1944.
18. Loof JP, Ploeger HD, van der Steur A, E.M. Meijers Instituut, editors. *The right to property: the influence of Article 1 Protocol no. 1 ECHR on several fields of domestic law*. Shaker Pub; 2000.
19. Jacobs HM. Private property and human rights: a mismatch in the 21st century?: private property and human rights. *Int J Soc Welf* 2013;**22**:S85–S101. https://doi.org/10.1111/ijsw.12044.
20. Ingram GK, Hong Y, Lincoln Institute of Land Policy, editors. *Property rights and land policies*. Lincoln Institute of Land Policy; 2009.
21. Nikolic D. Climate change and property rights changes. In: van Straalen F, Hartmann T, Sheehan J, editors. *Property rights and climate change: land use under changing environmental conditions*. Routledge; 2018.
22. Ricardo D. *On the principles of political economy and taxation*. Liberty Fund; 2004.
23. Norton RK, Meadows GA. Land and water governance on the shores of the Laurentian Great Lakes. *Water Int* 2014;**39**(6):901–20. https://doi.org/10.1080/02508060.2014.954661.

24. Suykens CBR, Tarlock D, Priest SJ, Doorn-Hoekveld WJ, van Rijswick HFMW. Sticks and carrots for reducing property-level risks from floods: an EU–US comparative perspective. *Water Int* 2019;**44**(5):622–39. https://doi.org/10.1080/02508060.2019.1640957.
25. Dai L, van Doorn-Hoekveld WJ, Wang RY, van Rijswick HFMW. Dealing with distributional effects of flood risk management in China: compensation mechanisms in flood retention areas. *Water Int* 2019;**44**(5):607–21. https://doi.org/10.1080/02508060.2019.1620014.
26. Sand M. Costs and benefits: why economic quantification in hazard mitigation policy threatens culture in coastal Louisiana. In: van Straalen F, Hartmann T, Sheehan J, editors. *Property rights and climate change: land use under changing environmental conditions*. Routledge; 2018.
27. Mishra S, Mazumdar S, Suar D. Place attachment and flood preparedness. *J Environ Psychol* 2010;**30**(2):187–97. https://doi.org/10.1016/j.jenvp.2009.11.005.
28. Otsuki K, Jasaw GS. redistribution of property rights in response to climate change in Ghana, West Africa. In: van Straalen F, Hartmann T, Sheehan J, editors. *Property rights and climate change: land use under changing environmental conditions*. Routledge; 2018.
29. Coase RH. The problem of social cost. *J Law Econ* 1960;**3**:1–44. http://www.jstor.org/stable/724810.
30. Slavíková L, Hartmann T, Thaler T. Financial schemes for resilient flood recovery. *Environ Hazards* 2020;**19**(3):223–7. https://doi.org/10.1080/17477891.2019.1703624.
31. Davy B. Polyrational property: rules for the many uses of land. *Int J Commons* 2014;**8**(2):472. https://doi.org/10.18352/ijc.455.
32. Hartmann T. *Clumsy floodplains. Responsive land policy for extreme floods*. Ashgate; 2011.
33. Slavíková L, Hartmann T, Thaler T. Paradoxes of financial schemes for resilient flood recovery of households. *WIREs Water* 2021;**8**(2):e1497. https://doi.org/10.1002/wat2.1497.
34. Berg M. *Was ist ein Schaden? Zur normativen Dimension des Schadensbegriffs in der Risikowissenschaft*. Verlag der Fachvereine; 1994.
35. Potočki K, Hartmann T, Slavíková L, Collentine D, Veidemane K, Raška P, et al. Land policy for flood risk management—toward a new working paradigm. *Earth's Future* 2022;**10**(4):e2021EF002491. https://doi.org/10.1029/2021EF002491.
36. Koelman M, Hartmann T, Spit TJM. It's not all about the money—landowner motivation and high voltage grid development. *J Environ Policy Plan* 2022;1–14. https://doi.org/10.1080/1523908X.2022.2093175.
37. Cairney P. Complexity theory in political science and public policy. *Political Stud Rev* 2012;**10**(3):346–58. https://doi.org/10.1111/j.1478-9302.2012.00270.x.
38. de Roo G. Spatial planning, complexity and a world 'out of equilibrium': Outline of a non-linear approach to planning. In: *Complexity and planning: systems, assemblages and simulations*. Ashgate Publishing. 2012. p. 141–75.
39. Douglas M. Four cultures: the evolution of a parsimonious model. *GeoJournal* 1999;**47**(3):411–5.
40. Hartmann T. Wicked problems and clumsy solutions: planning as expectation management. *Plan Theory* 2012;**11**(3):242–56. https://doi.org/10.1177/1473095212440427.

CHAPTER TWO

Urban sprawl: Theory and practice

Vito Imbrenda[a,*], Rosa Coluzzi[a], Leonardo Bianchini[b], Valerio Di Stefano[b], and Luca Salvati[c]

[a]Institute of Methodologies for Environmental Analysis - National Research Council of Italy (IMAA-CNR), Tito Scalo, PZ, Italy
[b]Department DAFNE, University of Tuscia, Viterbo, Italy
[c]Department of Economics and Law, University of Macerata, Macerata, Italy
*Corresponding author: e-mail address: vito.imbrenda@imaa.cnr.it

Contents

Abstract

Urban sprawl is among the most debated topics in the field of urbanism, regional science, ecology, economics, and geography. As this process involves different analysis' dimensions, sprawl is fascinating on the one side, but quite difficult to investigate on the other side. Having the objective to review the nature, dynamics, and consequences that low-density urban expansion is having on biophysical environments, this contribution provides a brief picture on the relationship between urban sprawl and environment quality with a focus on Europe. It is assumed that environmental change and urban transitions are progressively threatening peri-urban and rural landscape leading to worse socio-ecological conditions hard to manage with traditional planning instruments. Literature supports the idea that environmental policy and spatial planning should cope more effectively with the increasing vulnerability of "sprawling" urban regions to natural hazards.

Keywords: Environmental quality, Landscape, Soil consumption, Indicators, Europe

Advances in Chemical Pollution, Environmental Management and Protection, Volume 8
ISSN 2468-9289
https://doi.org/10.1016/bs.apmp.2022.10.017

1. Introduction

While urbanization has been closely associated with economic growth, demographic dynamics and social change, heterogeneous patterns and processes of regional development reflect the uneven spatial distribution of urbanization, the subtle impact of population trends and the consequent implications for resource management and environmental sustainability [1–6]. Differential urban growth mechanisms in advanced countries are assumed to reflect regional divides in both economic and environmental dimensions [7–13]. To better understand the impacts of such dynamics on land-use change, interdisciplinary research should better link socio-demographic and economic patterns with landscape transformations, urbanization and territorial processes of change (see [14-16], Fig. 1).

Compactness and dispersion were and are the ways with which cities have evolved [17]. There has always been an academic debate concerning both positive and negative traits of these two urbanization patterns [18–20]. However, only since the beginning of the 20th century, when world population rapidly increased and issues regarding the conservation

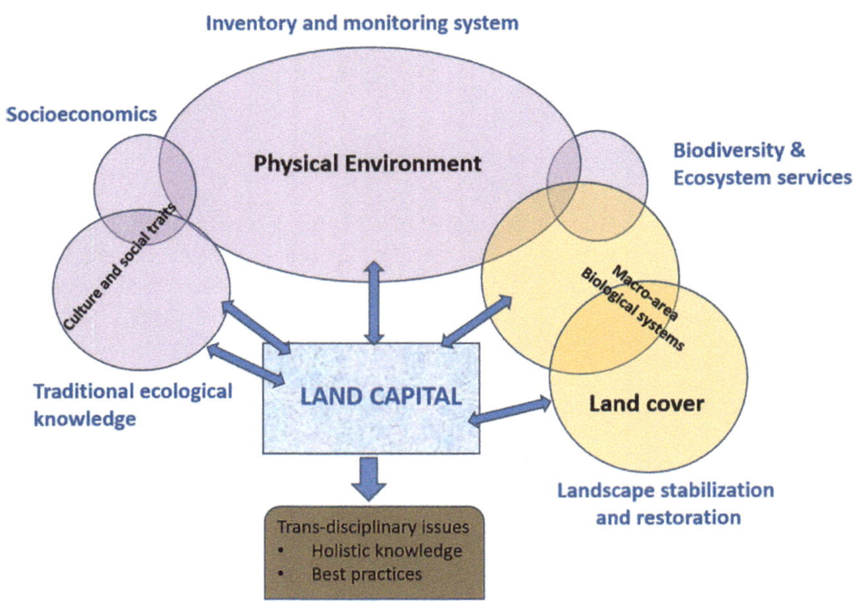

Fig. 1 A schematic description of the most relevant socioeconomic forces affecting land capital and thus influencing urbanization mechanisms and dynamics.

of the natural capital grew [21], this *querelle* on compact versus sprawl trajectories of urban development has become a matter of intense research [22–24]. Urban sprawl, i.e., the phenomenon of low-density settlement diffusion over large peri-urban areas, is amongst the most debated topics in the fields of urbanism with a huge involvement of other research areas, including environmental sciences, ecology, economics, geography, and sociology [25-29]. Because speaking of sprawl phenomena implies the involvement of different subjects of study, this process is extremely fascinating, but, at the same time, rather difficult to investigate because of the complex interplay between socioeconomic and ecological dimensions [30–32]. For these reasons, sprawl has and is attracting the scientific interest of much researchers from all over the world, having the aim of understanding nature, dynamics and consequences of this phenomenon on the natural capital [33–37]. Typically, urban sprawl was initially defined in reference to the rapid low-density outward expansion of United States cities [38], steaming back to the first half of the 20th century, fueled by the unstoppable growth of private car ownership and the preference for detached houses with gardens [39]. The first investigations of sprawl patterns and processes have been carried out in North America, where firstly the phenomenon appeared with heavy intensity ([40], see Fig. 2). Today, urban sprawl is amongst the major concerns in both advanced and emerging economies for its detrimental environmental impacts [41,42]. Nevertheless, the large majority of secondary sources regarding dispersed urbanization are still of the North American origin [43].

The aim of this work is to briefly discuss about the role held by urban dynamics and sprawling processes and to comment on their relationships with environmental quality and natural capital depletion [44]. After having

Fig. 2 Spatial configuration of Las Vegas in 1984 (on the left) and 2009 (on the right) illustrated through Landsat images and showing a huge expansion of urban areas over a relatively short time interval.

provided a general overview on the genesis of urbanization phenomena (e.g., [45]), we illustrate the main typical features and the classical morphological forms with which the process manifests itself [46]. Single-use zoning, low-density development, ribbon and leapfrog development, land-sealing of agroforestry areas, infrastructure-driven development and car-dependent communities have been described and interpreted as different typologies of urban sprawl [47]. In the subsequent sections, we analyze the impact and the environmental consequences of sprawl in qualitative and quantitative terms. The latter analysis has revealed particularly complex, because the general effects of ecosystems' worsening and natural resource consumption, linked to low-density urbanization, are not easy to quantify [48]. Nevertheless, for a better understanding of the environmental impact of sprawl, the contribution has focused on the energy and natural resources consumption differentials between compact and diffused patterns of urbanization [49].

Current literature highlights the integration of ecology with human demography, behavior, and socioeconomics as a relevant approach to capture and correctly manage ecological patterns and processes [50–52]. With a specific focus on Europe, our work encourages studies within a multidisciplinary arena, stimulating further literature reviews aimed at discussing these critical issues - proposing new theoretical frameworks, with empirical approaches, comparative works and case studies providing the necessary, informed ground to science and policy [35]. Soil and water consumption, raw materials growing demand (mainly for concrete) and energy consumption (fossil fuels for transport) are possible phenomena to be considered within the "sprawl" issue (e.g. [53]). Furthermore, this study should indicate the connection between sprawl, climate change and the increasing vulnerability of urban areas to extreme events, e.g., heat waves, flooding, earthquakes, just to mention the most relevant [24,54-56]. By extending urbanized areas over the neighboring semi-natural areas, probability and frequency of disasters linked to extreme weather events raises.

2. The morphological traits of sprawl

According to the geographical location where the phenomenon has been analyzed (e.g. North America or Europe, China or Latin America), scholars have postulated different definitions of urban sprawl, assuming different features, dynamics, effects, and impacts based on the nature of human societies [31]. Consequently, searching for a unique definition of sprawl is even more difficult when the differences in urbanization patterns and processes found in various socio-economic and environmental contexts are considered [57]. In this perspective, literature converges on comprehensive definitions of sprawl as a spatial pattern of urbanization associated with design features that encourage car dependency [32]. Nonetheless, other definitions of sprawl highlight different urban characteristics and dynamics, as we can see from the following quotations:

> "Urban sprawl refers to a gluttonous use of land, uninterrupted monotonous development, leapfrog discontinuous development and inefficient land use." *[58].*
> "Urban sprawl is a pattern of urban and metropolitan growth that reflects low density, automobile-dependent, exclusionary new development on the fringe of settled areas often surrounding a deteriorating city." *[59].*
> "Urban sprawl is random unplanned growth characterized by inadequate accessibility to essential land uses such as housing, jobs, and public services like schools, hospitals, and mass transit." *[60].*

It is possible to state that the term *"sprawl"* has a prevalent negative sense; social scientists generally depict sprawl as a problem [50]. Some do it explicitly by definition, others link it to detrimental consequences such as the decline of central cities or the worsening of public health conditions [45,61]. Neutral terms (such as *urban deconcentration, suburban expansion, counter-urbanization*) have not caught on, because these concepts do not distinguish between sprawl and suburbanization in general [54]. In other words, sprawl is frequently seen as "*a particular form of suburbanisation with several characteristics that differentiate it from other conceivable forms of suburbanisation.*"

From the cited definitions, it is possible to observe that even though there is still no consensus on the exact meaning of the word "sprawl," current literature has identified the key components of the concept [26]. For instance, Downs [62] stressed six features of sprawl, some of which distinguish it from other forms of suburbanization: (i) no limits placed on the outward suburban expansion; (ii) legal control over land-use, local services, transportation, property taxes, and fiscal policy divided among many small entities or jurisdictions, with no central agency responsible for the planning or control of these issues regionally; (iii) extensive "*leapfrog*" development; (iv) fragmented land ownership; (v) different types of land-use, spatially separated or zoned into distinct areas; and, finally, (vi) extensive strip commercial and residential development along larger suburban roads.

Downs' [62] formulation of sprawl is valuable because it suggests the need to think of and measure sprawl in terms of multiple indicators or dimensions (land-use, land price, housing characteristics, wealth and social profile of local communities) rather than simply in terms of low-density settlement patterns (see e.g., [63]). A report by Ewing *et al.* [64] gives a picture of some of the indicators suitable to estimate sprawl phenomena. In general, these indicators reflect the features outlined above and include: (i) low-density residential developments; (ii) a rigid separation of homes, services and workplaces; (iii) a network of roads marked by large blocks and poor access; and (iv) a lack of well-defined activity centers, such as "downtowns" and town centers (e.g. [26,51,54]). Finally, Ewing *et al.* [64] suggests that sprawl might be conceived as a "*process in which the spread of development across the landscape far outpaces population growth.*" In this sense, the definition given by Glaster [65] is the one that best facilitates the concept of sprawl as a process and not merely a spatial pattern. This process is triggered by social, economic and environmental drivers causing a decline in demand for land development in the center of the city whilst increasing it in fringe districts [55]. The city spreads over a larger surface (while the volume remains approximately constant). With this definition of sprawl, we must be very careful in distinguishing the (compact) growing process of the city with the spreading one.

3. Environmental implications of sprawl

Low-density urban patterns of development have many costs and implications. Most of these directly impact natural matrices and cause the worsening of social conditions, besides direct financial costs. In this section, analysis on the impact of sprawl will help us understanding whether anti-sprawl correctly boycott low-density urbanization in favor of compact growth. Sprawl impacts directly habitat functioning [66], but also has significant impacts on natural capital and produces socio-economic damages:

- wetlands, useful for flood control and wastewater treatment;
- forests, shrublands and grasslands involved in climate regulation;
- biodiversity factors that contribute to healthy and well-functioning ecosystems;
- ecosystems' services such as food production, water conservation, climate regulation, habitat support, cultural service; tourism industry damage by losing the scenic qualities of landscapes that attract tourists;
- social Fragmentation due to the isolation of families that cross the city only by car, losing the sense of place;
- The increasing levels of soil pollution, habitat disruption and fragmentation, degrade or destroy the environmental resources in the long term. The environmental impact of sprawl involves all the geographical scales

(i.e., local, regional and global; see [67]). An unintended consequence of low-density suburban growth is a larger resource consumption contributing to a greater environmental damage with respect to a compact development pattern. In this chapter, a comparison between the energy and natural resources consumption (which are directly related to impacts on the environment) of compact and low-density urbanization patterns is shown.

3.1 Land resource depletion

Suburbs *"are now the dominant residential, retail, and commercial centres of growth and political muscle"* and the continuation and replication of this trend *"place(s) enormous pressure on land, water, energy and other resources"*. Amongst the major concerns, regarding the sprawl process is that it *"eats into open space"* [44]. The U.S. EPA has created a *sprawl index* based on per capita consumption. According to a specific study conducted for 45 of the 50 most populous metropolitan regions in the US, a home owner whose year income is $50,000 living in a central area of the city has an average lot size of 7.8 thousand square feet (less than a quarter of an acre), while the average home owner in the suburbs with the same income has an average lot size of 12.3 thousand square feet [68]. Even if city-suburb land consumption differential varies among cities, the general trend is that suburbanites consume more land per capita. As suburban land grows, probably agricultural land declines [69]. The concept of sprawl conjures up images of concrete roads and parking lots replacing agricultural areas. As the population of urban areas grows, it is likely that land at the fringe of such areas will be transformed into urban land uses [70]. As land use/cover changes occur, consisting in the clearing of forests and the conversion of agricultural lands for urban development, quantity and quality of water supply are threatened [71,72]: as soil sealing rages over large areas, rainfall is less effectively absorbed and returned to groundwater. On the contrary, relatively more stormwater flows to streams and rivers is carried downstream. This phenomenon also occurs in the case of the snow-melt, especially early in the melting process. With less groundwater recharge, communities depending on groundwater for their drinking water may face shortages [73,74].

Agricultural production depends on a complex assortment of factors, such as the overall quality of the natural properties (soil fertility, soil moisture, solar energy, climate), the amount of inputs of human, animal, and fossil fuel energy via labor and machinery, other inputs and practices, (fertilizers, pesticides, irrigation, soil conservation practices), and agricultural development programs (see [75]). The impact of sprawl on agricultural activities is direct, by sealing previously cultivated lands and indirectly by expanding in fertile areas bordering the urban land uses. This phenomenon is enabled by the suitable features of the neighboring agricultural soils facilitating the building of commercial and residential areas. As a consequence, there is a strong competition for the use of these soils, with a prevalent conversion from agricultural to most profitable land uses (those providing the more immediate economic return, see Ref. [76]). To compete with alternative land uses, farmers in urbanized areas must opt for intensive agricultural practices or for more profitable crops.

Some researchers showed that since the mid-twentieth century, American farmers have been producing more crops on fewer acres; they also estimated that crop production increased with the use of agrochemicals with a major loss of farmland [77,78]. To prevent habitat loss, especially for endangered and threatened species, and enhance the recreational and aesthetic value of forests, the cutting of timber should be drastically reduced [79-82]. In fact, the expansion of residential land uses makes the presence of forests more valuable in terms of cultural services rather than fiber production [83]. Unavoidably, urban growth produces alterations of the landscapes by changing the prior land uses/covers or fragmenting them, reducing the size of the remaining habitat patches and their degree of connection [84,85]. The effects of the sprawl are both directly appreciable on the physical reduction of the involved habitats and indirectly on the adjacent habitats due to light and noise pollution coming from the neighboring areas [86,87]. This is confirmed by the National Wildlife Federation (NWF) [88] affirming: "*artificial lighting may also fragment the landscape and habitat for wildlife, even if there are connecting corridors.*"

Another consequence of the suburban and exurban development is the closure and/or re-location of the quarries (mineral extraction) far from the urban center. This can represent a serious issue for several reasons. Firstly, industrial minerals such as limes, sands, and gravels, used to produce cement and required in large amounts for building, are commodities very sensitive to transportation costs. Secondly, the shift of industrial operations to other sites can compromise or destroy ecological and aesthetic integrity of the remaining open areas. In the light of this, it is evident that sprawl leads to higher consumptions of fossil fuels and natural resources (agricultural lands and forests). This implies higher levels of air pollution, the decline of farmland activities, loss of wildlife habitats and issues connected with water supply (e.g., [89–92]). Other environmental impacts related to natural resource consumption include a lower water quality stemming from urban "non-point" sources of pollution; destabilization of stream channels and increasing probability of flooding events due to stormwater runoff from urbanized areas; alterations of local climates, including the urban heat island effect; loss and fragmentation of wildlife habitats; degradation of landscape aesthetics; and increasing evidence of noise and light pollution (see, e.g., [93–102]).

To understand the future pressure on urbanization and, possibly, urban diffusion, data from the World Urbanization Prospects have been considered. The overall picture provided by these data indicates that urban population is expected to significantly grow in the future [103] both in developed and developing countries.

This further puts in evidence the necessity of planned growth (for developing countries) and containment (for developed countries) strategies to tackle the pressure on sprawl in the perspective to achieve a real sustainable development [104–107].

4. A brief outlook on urban trends in Europe

Urban sprawl is advancing in many metropolitan areas of the world, and it is becoming a common issue in most of the European cities, irrespective of their geographical, socio-economic or political features. The rising interest amongst European countries in mapping and analyzing this specific pattern of growth is testified by the increasing number of studies and research projects (e.g., [108]). They focused prevalently on the spatial dimension of the phenomenon by increasingly using remote observations (e.g., satellite data, see [109]). By the late 1980s the first environmental issues concerning urban sprawl began to appear in the European social and political debates. By then, the control of sprawl had become a major consideration in the framework of urban policies in most European countries. The European Commission stated that: *"uncontrolled growth results in increased levels of private transport, increased energy consumption, makes infrastructures and services more costly and has negative effects on the quality of the countryside and the environment. (...) It is therefore necessary to work together to find sustainable solutions for planning and managing urban growth"* [39].

Starting in the late nineties the EU began to work on town planning strategies mainly focused on the achievement of mixed land-use and denser urban development in order to reduce the negative impact of sprawl on the social and environmental dimensions [110].

As in North America and other parts of the world, throughout Europe urban sprawl is becoming a consolidated threat [111]. The environmental, social and economic effects of the phenomenon for cities and countryside of Europe are becoming always more manifest and require urgent action, especially in the feared perspective of the escalation of climate change impacts [112]. For this reason, modern town planning should be developed with the aim of controlling urban expansion. Nevertheless, the first attempts of urban planning did not reach the proposed objective. For example, between 1922 and 1939 over 340,000 ha of rural land in England and Wales were converted to urban land uses (representing about a 40% increase in the total urban area of the country, see e.g., [113]). In the aftermath of World War II, many European countries invested heavily in planned urban expansions schemes. Most of these schemes produced peripheral extensions of existing core urban areas with very low densities. The unstoppable economic rise experienced by Western Europe in the post-war period triggered a considerable demographic growth [114]. Urbanization of agricultural areas was the unavoidable consequence of a bigger population and a growing economy [115]. Land change was quite sudden, as well as the transformation of urban landscapes across the continent. Whilst North-West European cities reached their urban peak towards the middle of the 20th century, most conurbations of South and East Europe followed increasing trends until nearly the end of the last century [116,117].

Another critical parameter for the effects of urban sprawl is connected to the changes occurred in the industrial structures. There is a number of trends that can be observed in Europe: the movement of production to other regions and countries (the so-called globalization); the decentralization of employment to suburban locations; the development of new forms of employment, especially in the service sector; the shrinkage and closure of traditional industries. The latter had the effect of removing employment and weakening the links between inner urban residential and workplaces areas. This way, outward migration of workers to suburban areas was encouraged. By the end of the 20th century, tackling urban sprawl was becoming a global affair. In 1992, the United Nation's Agenda 21 asked that all states promote sustainable patterns of land-use and development in order to contrast the sprawling processes [39].

The total population of the European Union in 2022 was just over 748 million, with an average population density of about 34 inhabitants per km^2 (https://www.worldometers.info/world-population/europe-population/#:~:text=Europe%20Population%20(LIVE)&text=The%20population%20density%20in%20Europe,87%20people%20per%20mi2) which is pretty much the same of the average of the United States (https://www.macrotrends.net/countries/USA/united-states/population-density#:~:

text=The%20current%20population%20density%20of,a%200.38%25%20 increase%20from%202021). Due to the ageing and low fertility rates of the Europeans, it is predicted that population will increase moderately and will depend mainly on inward migration form countries outside the continent [118]. Within Europe, constant migration from East to West and from rural to urban areas still seems to be the general trend [119]. The increase of European population living in urban areas means that urban areas must provide accommodation for millions of additional inhabitants. The pressure linked to urbanization processes is significant. Besides the increasing housing demand caused by urban migration, new and changing economic functions are requiring cities to create more peripheral land for commercial and industrial uses [120]. Moreover, this process is further reinforced by the competition for capital attraction amongst cities [121].

5. Quantifying the environmental impact of urban sprawl

In this section, the environmental impacts attributable to sprawl processes in the European countries will be discussed.

As clearly indicated in Fig. 3, representing past, present and future land-use changes according to the forest transition theory, urban land uses and forest expansion progressively have reduced and will continue to reduce the extent of agricultural areas in developed countries (see, e.g., [123–126]). The growth of European cities in recent years has primarily occurred on

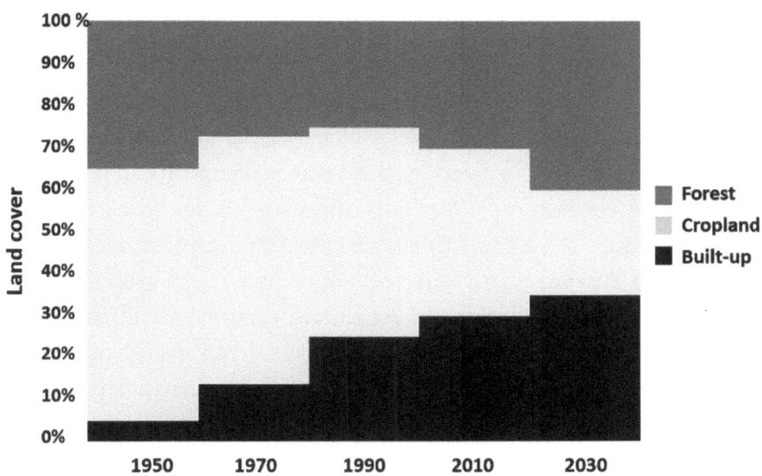

Fig. 3 A schematic representation of past, present and future land-use changes in developed countries according to the forest transition theory [122].

former agricultural lands (Fig. 3). Typically, forests are far from cities and therefore urban development and agriculture are competing for the same land, as agricultural lands, adjacent to existing urban areas, represent their ideal extension [127]. The motivations of farmers in these socio-economic dynamics are directed to capitalize financial benefits for the sale of farmland for the building of new housing or other urban land uses [128]. In Poland, for example, between the period 2004–2006 the price of agricultural land increased on average by 40% [129].

Being a non-renewable resource, soil should be protected, otherwise the loss of agricultural land will imply detrimental consequences on crop yields, food security and possible repercussions also on biodiversity with the loss of valuable biotopes for many species of animals, depending on the specific type of agriculture that was practiced [45,130-132]. In addition to this, the quality of the agricultural lands, still not urbanized but falling in the vicinity of sprawling cities, has also been reduced. All these typical impacts of sprawl are well represented in the landscapes of the Mediterranean coastal areas [133].

5.1 Urban growth and water resources

Land-use changes linked to urban expansion are also altering water/land-surface features. If the capacity of certain areas to preserve ecological and social benefits guaranteed by groundwater diminishes over time, this could lead to conflicts due to competition for water resources (see e.g., [134]). These conditions generally produce strong migratory flows of people looking for places offering a better quality of life [48]. Mediterranean regions of Europe, where desertification processes are active [7,135–143] appear particularly vulnerable to such a condition. Reduction of groundwater recharge could negatively impact also on the overall hydrological cycle, included that of wetlands surrounding sprawled cities. Moreover, urban diffusion in mountain areas of Europe is of particular interest, because these zones are commonly recognized as the "water reservoirs of Europe" and hosting valuable ecosystems. Currently, they are under threat from urban expansion (see, e.g., [144,145]). New and modern transport facilities enable the commuting toward many metropolitan areas with populations over 250,000 inhabitants that lie close to mountain regions, encouraging urbanization in the mountain zones [146]. Increased transit and tourist traffic, specifically day tourism from major cities, also add disturbance to the mountain ecosystems [147]. More balanced solutions are desirable in the urban-mountain relationship if the precious ecosystems of these areas must be conserved.

5.2 Raw material consumption

Environmental conflicts associated with sprawl phenomena include the expansion of quarries, often located in the neighboring of nature sites, and the over-extraction of gravel from river beds. Transport-related energy consumption in urban areas depends on a number of factors including the nature of the rail and road networks, the extent of the development of mass transportation systems, and the modal split between public and private transport. Evidence shows that there is a significant increase in travel-related energy consumption in cities as densities fall [148]. Fundamentally, the sprawling cities are dominated by a relatively energy inefficient car use, as cars are often the only practical alternative to more energy efficient public transportation systems, that are typically too expensive [149]. Increased transport-related energy consumption is in turn leading to an increase in the emission of CO_2 to the atmosphere. The relationship between population densities and CO_2 emissions is apparent as emissions increase progressively with the fall of urban densities. Although there are several factors explaining differentials in CO_2 emissions between cities, including levels of industrial activities and local climatic conditions, the predominance of car borne transportation in sprawling cities is evidently the major factor in the growth of urban green-house gas emissions [150]. Urban sprawl therefore poses significant threats to the achievement of goals of the Paris Agreement. Sprawl also impacts on the length of trips required to collect municipal waste for processing them at waste treatment plants placed at ever greater distances [151]. The material cycle is becoming geographically decoupled with increasing transport demands, impacting on transport-related energy consumption and pollution emissions.

5.3 Climate change

Sprawl related growth of urban transport and greenhouse gas emissions have critical repercussions on global warming, with the expectation of increasingly frequent and severe weather events [152,153] implying a net growth of natural disasters (e.g., river and coastal flooding, heatwave, frost see [154–156]). Risks deriving from the continuous development of urban areas coupled with a changing climate are the responsible for the major floods in Europe that have affected large urban populations (see e.g., [157]). Recent estimates for the World Health Organization (WHO) European Region report that in the time interval 2000–2014 more than 2000 people were killed by coastal and inland floods affecting about 8.7 million (see https://www.eea.europa.eu/data-and-maps/indicators/

floods-and-health-1/assessment). These expected transformations pose major challenges for urban planning clearly focused on the regulation of urban sprawl along the coastal fringes, as well as on the development of sprawling areas across greenfield sites in the river valleys and lowlands [158]. The flooding of the European coastal areas due to rising sea levels and climate change is particularly worrying if we consider the concentration of urban populations along the coasts and the impact of tourism activities in these areas [159,160]. The European countries appearing more prone to coastal flooding include the Netherlands and Belgium, where more than 85% of the coastal areas is under 5 m elevation [161,162]. Other countries at risk include Germany and Romania where 50% of the coastline is below 5 m, Poland (30%) and Denmark (22%), as well as France, the United Kingdom and Estonia where lowlands cover 10–15% of the country [163]. Overall, 9% of all European coastal areas lie below 5 m elevation [164]. Even considering conservative estimates of sea level rise, a significant part of the European population living in coastal zones is highly vulnerable to sea level rise and flooding [165]. This fact is not a specific concern directly generated by sprawl, however, the management of these risks and the planning for adaptation will be made more complex if sprawl phenomena are not controlled.

5.4 Impact on society and urban quality

Changes in lifestyle associated with sprawl contribute to the increase of the demand of natural resources [166]. People are increasingly living in individual households, reflecting a tendency towards less efficient use of the spaces and thus requiring more resources per capita than larger households [167]. For instance, a two-person household uses 300 L of water per day, two single households use 210 L each. A two-person household will use 20% less energy consumption than two single person households [168]. The number of households grew by 11% between 1990 and 2000, a trend acting as a driver of urban expansion [169]. The general trend is for greater consumption of resources per capita with a corresponding growth in environmental impacts. This condition amplifies the effects of the long-standing deficit of the large European cities (about 60%) in terms of over-exploitation of their groundwater resources [170]. Moreover, market-oriented land-use allocations driving urban expansion and the transformation of economic activity often result in the partial or total abandonment of former industrial areas [171]. As a consequence, there are many derelict or underused industrial zones throughout Europe located in the peripheral areas of urban agglomerations [172]. For example, in Spain about 50% of sites contaminated from past industrial activities are located in urban areas, and in Austria it is estimated that abandoned industrial sites cover about 2% of all urban areas [173]). Generally, the efficiency savings of more compact models of city development as compared with market-driven suburbanization can be as high as 20–45% in land resources, 15–25% in the building of local roads and 7–15% savings in the provision of water and sewage facilities [174,175]. Lastly, urban sprawl generates many adverse environmental impacts that have direct effects on the quality of life and human health in cities, such as poor air quality and high noise levels that often exceed the established safety limits [176]. It is estimated that approximately 20 million Europeans suffer from respiratory problems linked to air pollution [177].

6. Concluding remarks

What is urban sprawl and how does it adapt to the different territorial contexts in which it is taking place? To answer this question, both formal and informal definitions of sprawl have been extensively investigated. The former definitions illustrate different features of sprawl, but at the same time they have evidenced the lack of a global and unique concept of the urban process being discussed. This goes in the direction of the initial hypothesis of this research, that is the necessity of analyzing sprawl through comparative analysis, as the phenomenon shows different features and effects according to the specific territorial context investigated. Informal definitions have been also discussed, as these appear useful in giving a solid comprehension on the concept of exurban development. The sand-castle metaphor remains efficient in highlighting the main feature of sprawling patterns of urbanization: the "volume" of the city can be considered approximately constant, but it is spread over a larger surface. Therefore, the concept of sprawl has been associated to the transformation process of the density gradient line of urban and residential activities, which will result always less steep as the phenomenon takes place. To emphasize the importance of the process of transformation rather than the final urban configuration of sprawling cities, the necessity of treating the phenomenon as a verb more than a noun and differentiating it from the common forms of the urban growth process has been underlined.

Besides the general impact of sprawl, littoralization along the European shores, taken as a paradigmatic example of urban diffusion, implies a loss of agricultural and natural land (which is amongst the richest and most productive of the region), the destruction of highly valuable natural habitats, the degradation and pollution of shores and internal/open waters, the reduction of small-scale fishing and the increasing vulnerability of these areas to extreme events. With regards to this last point, research shows that the continuous pressure on environmental matrices, due to human-induced demand factors (urbanization and natural resources consumption), is increasing the disaster potential of cities and megacities. The growing vulnerability of urban centers is testified by the increasing damages related to natural hazards. This evidence supports the idea that national policy and regional planning should cope more effectively with the increasing vulnerability of large urban regions to natural hazards.

References

1. Bianchini L, Egidi G, Alhuseen A, Sateriano A, Cividino S, Clemente M, et al. Toward a dualistic growth? Population increase and land-use change in Rome, Italy. *Land* 2021;**10**(7):749.
2. Imbrenda V, Quaranta G, Salvia R, Egidi G, Salvati L, Prokopovà M, et al. Land degradation and metropolitan expansion in a peri-urban environment. *Geomat Nat Haz Risk* 2021;**12**:1797–818.
3. Salvati L, Sateriano A, Grigoriadis E. Crisis and the city: profiling urban growth under economic expansion and stagnation. *Lett Spat Resour Sci* 2016;**9**(3):329–42.
4. Carlucci M, Chelli FM, Salvati L. Toward a new cycle: short-term population dynamics, gentrification, and re-urbanization of Milan (Italy). *Sustainability* 2018;**10**(9):3014.
5. Salvati L, Zitti M. Territorial disparities, natural resource distribution, and land degradation: a case study in southern Europe. *GeoJournal* 2007;**70**(2):185–94.
6. Salvati L, Zitti M. Substitutability and weighting of ecological and economic indicators: exploring the importance of various components of a synthetic index. *Ecol Econ* 2009;**68**(4):1093–9.
7. Lanfredi M, Egidi G, Bianchini L, Salvati L. One size does not fit all: a tale of polycentric development and land degradation in Italy. *Ecol Econ* 2022;**192**:107256. https://doi.org/10.1016/j.ecolecon.2021.107256.
8. Salvati L, Zitti M, Carlucci M. In-between regional disparities and spatial heterogeneity: a multivariate analysis of territorial divides in Italy. *J Environ Plan Manag* 2017;**60**(6):997–1015.
9. Salvia R, Halbac-Cotoara-Zamfir R, Cividino S, Salvati L, Quaranta G. From rural spaces to peri-urban districts: metropolitan growth, sparse settlements and demographic dynamics in a Mediterranean region. *Landarzt* 2020;**9**(6):200.
10. Ciommi M, Chelli FM, Carlucci M, Salvati L. Urban growth and demographic dynamics in southern Europe: toward a new statistical approach to regional science. *Sustainability* 2018;**10**(8):2765.
11. Ciommi M, Chelli FM, Salvati L. Integrating parametric and non-parametric multivariate analysis of urban growth and commuting patterns in a European metropolitan area. *Qual Quant* 2019;**53**(2):957–79.
12. Salvati L. Towards a polycentric region? The socio-economic trajectory of Rome, an 'eternally Mediterranean' city. *Tijdschr Econ Soc Geogr* 2014;**105**(3):268–84.

13. Schneider A, Woodcock CE. Compact, dispersed, fragmented, extensive? A comparison of urban growth in 25 global cities using remotely sensed data, pattern metrics and census information. *Urban Stud* 2008;**45**:659–92.
14. Egidi G, Salvati L, Falcone A, Quaranta G, Salvia R, Vcelakova R, et al. Re-framing the latent nexus between land-use change, urbanization and demographic transitions in advanced economies. *Sustainability* 2021;**13**(2):533.
15. Gavalas VS, Rontos K, Salvati L. Who becomes an unwed mother in Greece? Sociodemographic and geographical aspects of an emerging phenomenon. *Popul Space Place* 2014;**20**(3):250–63.
16. Legget RF. *Cities and geology*. New York: McGraw-Hill; 1973.
17. Rubiera-Morollón F, Garrido-Yserte R. Recent literature about urban sprawl: a renewed relevance of the phenomenon from the perspective of environmental sustainability. *Sustainability* 2020;**12**(16):6551.
18. Bueno-Suárez C, Coq-Huelva D. Sustaining what is unsustainable: a review of urban sprawl and urban socio-environmental policies in North America and Western Europe. *Sustainability* 2020;**12**(11):4445.
19. Colsaet A, Laurans Y, Levrel H. What drives land take and urban land expansion? A systematic review. *Land Use Policy* 2018;**79**:339–49.
20. Hosseini SH, Hajilou M. Drivers of urban sprawl in urban areas of Iran. *Pap Reg Sci* 2019;**98**(2):1137–58.
21. Ismael HM. Urban form study: the sprawling city—review of methods of studying urban sprawl. *GeoJournal* 2021;**86**(4):1785–96.
22. Bruegmann R. *Sprawl: a compact history*. Chicago: University of Chicago Press; 2005.
23. Di Feliciantonio C, Salvati L. 'Southern' alternatives of urban diffusion: investigating settlement characteristics and socio-economic patterns in three Mediterranean regions. *Tijdschr Econ Soc Geogr* 2015;**106**(4):453–70.
24. Feng Q, Gauthier P. Untangling urban sprawl and climate change: a review of the literature on physical planning and transportation drivers. *Atmosfera* 2021;**12**(5):547.
25. Nechyba TJ, Walsh RP. Urban sprawl. *J Econ Perspect* 2004;**18**(4):177–200.
26. Wang X, Shi R, Zhou Y. Dynamics of urban sprawl and sustainable development in China. *Socio Econ Plan Sci* 2020;**70**:100736.
27. Zambon I, Colantoni A, Salvati L. Horizontal vs vertical growth: understanding latent patterns of urban expansion in large metropolitan regions. *Sci Total Environ* 2019; **654**:778–85.
28. Duvernoy I, Zambon I, Sateriano A, Salvati L. Pictures from the other side of the fringe: urban growth and peri-urban agriculture in a post-industrial city (Toulouse, France). *J Rural Stud* 2018;**57**:25–35.
29. Perrin C, Nougarèdes B, Sini L, Branduini P, Salvati L. Governance changes in peri-urban farmland protection following decentralisation: a comparison between Montpellier (France) and Rome (Italy). *Land Use Policy* 2018;**70**:535–46.
30. Davoudi S. Polycentricity in European spatial planning: from an analytical tool to a normative agenda. *Eur Plan Stud* 2003;**11**(8):979–99.
31. Han J. Can urban sprawl be the cause of environmental deterioration? Based on the provincial panel data in China. *Environ Res* 2020;**189**:109954.
32. Mouratidis K. Compact city, urban sprawl, and subjective well-being. *Cities* 2019; **92**:261–72.
33. Chin N. *Unearthing the roots of urban sprawl: a critical analysis of form, function and methodology*. vol. 47. Centre for Advanced Spatial Analysis of the University College London; 2002. Working Paper, Series.
34. Hasse JE. A geospatial approach to measuring new development for characteristics of sprawl. *Landsc J Des Plan Manage Land* 2004;**23**:52–67.

35. Hasse JE, Lathrop RG. Land resource impact indicators of urban sprawl. *Appl Geogr* 2003;**23**:159–75.
36. Salvati L, Gitas I, Di Giacomo TV, Saradakou E, Carlucci M. Sprawl matters: the evolution of fringe land, natural amenities and disposable income in a Mediterranean urban area. *Environ Dev Sustain* 2017;**19**(2):727–43.
37. Xiaojun Y. *Urban remote sensing: monitoring, synthesis and modeling in the urban environment*. Chichester: John Wiley & Sons; 2011.
38. Brueckner JK. Urban sprawl: diagnosis and remedies. *Int Reg Sci Rev* 2000;**23**(2):160–71.
39. European Environmental Agency. *Urban sprawl in europe—the ignored challenge*; 2006. Copenhagen: EEA Report no. 10, Office for Official Publications of the European Communities, ISBN 92-9167-887-2, ISSN 1725-9177.
40. Downs A. Some realities about sprawl and urban decline. *Hous Policy Debate* 1999;**10** (4):955–74.
41. Frenkel A, Ashkenazi M. The integrated sprawl index: measuring the urban landscape in Israel. *Ann Reg Sci* 2007;**42**(1):99–121.
42. Di Feliciantonio C, Salvati L, Sarantakou E, Rontos K. Class diversification, economic growth and urban sprawl: evidences from a pre-crisis European city. *Qual Quant* 2018;**52**(4):1501–22.
43. Johnson MP. Environmental impacts of urban sprawl: a survey of the literature and proposed research agenda. *Environ Plan B* 2001;**33**:717–35.
44. Kahn ME. The environmental impact of suburbanization. *J Policy Anal Manage* 2000;**19** (4):569–86.
45. Hatab AA, Cavinato MER, Lindemer A, Lagerkvist CJ. Urban sprawl, food security and agricultural systems in developing countries: a systematic review of the literature. *Cities* 2019;**94**:129–42.
46. Hall P. The future of metropolis and its form. *Reg Stud* 1997;**31**(3):211–20.
47. Jaret C, Ghadge R, Reid LW, Adelman RM. The measurement of suburban sprawl: an evaluation. *City Community* 2009;**8**(1):65–84.
48. Craglia M, Leontidou L, Nuvolati G, Schweikart J. Towards the development of quality indicators in the 'digital' city. *Environ Plan B* 2004;**31**:51–64.
49. Frenkel A, Ashkenazi M. Measuring urban sprawl: how can we deal with it? *Environ Plan B* 2008;**35**:56–79.
50. Li G, Li F. Urban sprawl in China: differences and socioeconomic drivers. *Sci Total Environ* 2019;**673**:367–77.
51. Masoumi HE, Hosseini M, Gouda AA. Drivers of urban sprawl in two large Middle-eastern countries: literature on Iran and Egypt. *Hum Geogr* 2018;**12**(1):55–79.
52. Zambon I, Salvati L. Metropolitan growth, urban cycles and housing in a Mediterranean country, 1910s–2010s. *Cities* 2019;**95**:102412.
53. Haase D, Nuissl H. The urban-to-rural gradient of land use change and impervious cover: a long-term trajectory for the city of Leipzig. *Land Use Sci* 2010;**5** (2):123–42.
54. Bosch M, Chenal J, Joost S. Addressing urban sprawl from the complexity sciences. *Urban Sci* 2019;**3**(2):60.
55. Koprowska K, Łaszkiewicz E, Kronenberg J. Is urban sprawl linked to green space availability? *Ecol Indic* 2020;**108**:105723.
56. Turok I, Mykhnenko V. The trajectories of European cities, 1960-2005. *Cities* 2007;**24** (3):165–82.
57. Longhi C, Musolesi A. European cities in the process of economic integration: towards structural convergence. *Ann Reg Sci* 2007;**41**:333–51.
58. Peiser R. Decomposing urban sprawl. *Town Plan Rev* 2001;**3**(72):96–108.
59. Squires GD. *Urban sprawl: causes, consequences, and policy decisions*. Washington DC: Urban Institute Press; 2002.
60. Bullard RD, Glenn SJ, Torres AO. *Sprawl city*. Washington DC: Island Press; 2000.

61. Frumkin H. *Urban sprawl and public health public health reports*. Vol. 117–216. Association of Schools of Public Health; 2002.
62. Downs A. The challenge of our declining big cities. *Hous Policy Debate* 1997;**8**:359–408.
63. Salvati L, Ciommi MT, Serra P, Chelli FM. Exploring the spatial structure of housing prices under economic expansion and stagnation: the role of socio-demographic factors in metropolitan Rome, Italy. *Land Use Policy* 2019;**81**:143–52.
64. Ewing R, Pendall R, Chen D. *Measuring sprawl and its impact*. Washington, DC: Smart Growth America; 2002.
65. Glaster G, Hanson R, Ratcliffe MR, Wolman H, Coleman S, Freihage J. Wrestling sprawl to the ground: defining and measuring an elusive concept. *Hous Policy Debate* 2001;**12**(4):681–717.
66. McInnes R. Urban development, biodiversity and wetland management, export workshop 16–17 November 2009. *Oxford Unhabitat* 2010;1–57.
67. Barnes KB, Morgan JM, Roberge MC, Lowe S. Sprawl development: its patterns, consequences, and measurement. Technical Report-January 2002, Geospatial Research and Education Laboratory. *Department of Geography and Environmental Planning*, Baltimore, Maryland, USA: Towson University; 2002.
68. Ewing R, Pendall R, Chen D. Measuring sprawl and its transportation impacts. *Transp Res Rec* 2003;**1831**(1):175–83.
69. Zambon I, Colantoni A, Carlucci M, Morrow N, Sateriano A, Salvati L. Land quality, sustainable development and environmental degradation in agricultural districts: a computational approach based on entropy indexes. *Environ Impact Assess Rev* 2017;**64**:37–46.
70. Salvati L, Rontos K, Morelli VG. Urban sprawl and implications for sustainable transportation: Analysing changing commuting patterns in a Mediterranean city region. *Int J Sustain Soc* 2015;**7**(4):333–52.
71. Chelleri L, Schuetze T, Salvati L. Integrating resilience with urban sustainability in neglected neighborhoods: challenges and opportunities of transitioning to decentralized water management in Mexico City. *Habitat Int* 2015;**48**:122–30.
72. Salvati L. Agro-forest landscape and the 'fringe'city': a multivariate assessment of land-use changes in a sprawling region and implications for planning. *Sci Total Environ* 2014;**490**:715–23.
73. Bouziotas D, Rozos E, Makropoulos C. Water and the city: exploring links between urban growth and water demand management. *J Hydroinf* 2015;**17**(2):176–92.
74. Löwe R, Mair M, Pedersen AN, Kleidorfer M, Rauch W, Arnbjerg-Nielsen K. Impacts of urban development on urban water management—limits of predictability. *Comput Environ Urban Syst* 2020;**84**:101546.
75. Simoniello T, Coluzzi R, D'Emilio M, Imbrenda V, Salvati L, Sinisi R, et al. Going conservative or 790 conventional? investigating farm management strategies in between economic and environmental sustainability in 791 Southern Italy. *Agronomy* 2022;**12**:597. https://doi.org/10.3390/agronomy12030597.
76. Zambon I, Benedetti A, Ferrara C, Salvati L. Soil matters? A multivariate analysis of socioeconomic constraints to urban expansion in Mediterranean Europe. *Ecol Econ* 2018;**146**:173–83.
77. Furuseth OJ, Pierce JT. *Agricultural land in an urban society*. Washington, DC: Association of American Geographers; 1982.
78. Biasi R, Brunori E, Smiraglia D, Salvati L. Linking traditional tree-crop landscapes and agro-biodiversity in Central Italy using a database of typical and traditional products: a multiple risk assessment through a data mining analysis. *Biodivers Conserv* 2015;**24**(12):3009–31.
79. Corona P. The culture of the forest. *Ann Silvicult Res* 2022;**47**(1):54.
80. Nocentini S, Buttoud G, Ciancio O, Corona P. Managing forests in a changing world: the need for a systemic approach. A review. *For Syst* 2017;**26**(1):eR01-eR01.

81. Bajocco S, Dragoz E, Gitas I, Smiraglia D, Salvati L, Ricotta C. Mapping forest fuels through vegetation phenology: the role of coarse-resolution satellite time-series. *PLoS One* 2015;**10**(3):e0119811.
82. Recanatesi F, Clemente M, Grigoriadis E, Ranalli F, Zitti M, Salvati L. A fifty-year sustainability assessment of Italian agro-forest districts. *Sustainability* 2016;**8**(1):32.
83. Tomao A, Quatrini V, Corona P, Ferrara A, Lafortezza R, Salvati L. Resilient landscapes in Mediterranean urban areas: understanding factors influencing forest trends. *Environ Res* 2017;**156**:1–9.
84. Barlow SA, Munn IA, Cleaves DA, Evans DL. The effects of urban sprawl on timber harvest. *J Forestry* 1998;10–4.
85. Nickayin SS, Salvati L, Coluzzi R, Lanfredi M, Halbac-Cotoara-Zamfir R, Salvia R, et al. What happens in the city when long-term urban expansion and (un)sustainable fringe development occur: the case study of Rome. *ISPRS Int J Geo Inf* 2021;**10**(4):231.
86. Lamphar H. Spatio-temporal association of light pollution and urban sprawl using remote sensing imagery and GIS: a simple method based in Otsu's algorithm. *J Quant Spectrosc Radiat Transf* 2020;**251**:107060.
87. McCoshum SM, Geber MA. Land conversion for solar facilities and urban sprawl in southwest deserts causes different amounts of habitat loss for Ashmeadiella bees. *J Kansas Entomol Soc* 2020;**92**(2):468–78.
88. National Wildlife Federation (NWF). *Smart growth and wildlife*; 2001. https://www.nwf.org/Magazines/National-Wildlife/2001/Action-Report-Oct-Nov-2001. [Accessed 20 October 2022].
89. Aguilar AG. Peri-urbanization, illegal settlements and environmental impact in Mexico City. *Cities* 2008;**25**:133–45.
90. Alphan H. Land use change and urbanisation of Adana, Turkey. *Land Degr Develop* 2003;**14**:575–86.
91. Attorre F, Rossetti A, Sbrega B, Bruno F. Landscape changes in Rome, Italy. *Coenosis* 1998;**13**:564–78.
92. De Marco A, Proietti C, Anav A, Ciancarella L, D'Elia I, Fares S, et al. Impacts of air pollution on human and ecosystem health, and implications for the National Emission Ceilings Directive: insights from Italy. *Environ Int* 2019;**125**:320–33.
93. Carneiro E, Lopes W, Espindola G. Linking urban sprawl and surface urban heat island in the Teresina–Timon conurbation area in Brazil. *Landarzt* 2021;**10**(5):516.
94. Cecchini M, Zambon I, Pontrandolfi A, Turco R, Colantoni A, Mavrakis A, et al. Urban sprawl and the 'olive' landscape: sustainable land management for 'crisis' cities. *GeoJournal* 2019;**84**(1):237–55.
95. D'Emilio M, Coluzzi R, Macchiato M, Imbrenda V, Ragosta M, Sabia S, et al. Satellite data and soil magnetic susceptibility measurements for heavy metals monitoring: findings from Agri Valley (Southern Italy). *Environ Earth Sci* 2018;**77**:63. https://doi.org/10.1007/s12665-017-7206-4.
96. Franci F, Mandanici E, Bitelli G. Remote sensing analysis for flood risk management in urban sprawl contexts. *Geomat Nat Haz Risk* 2015;**6**(5–7):583–99.
97. McMahan JB, Weber KT, Sauder JD. Using remotely sensed data in urban sprawl and green space analyses. *Int J Therm Sci* 2002;**8**(1):30–7.
98. Radeloff VC, Hammer RB, Stewart SI. Rural and suburban sprawl in the US Midwest from 1940 to 2000 and its relation to forest fragmentation. *Conserv Biol* 2005;**19**(3):793–805.
99. Tu J, Xia ZG, Clarke KC, Frei A. Impact of urban sprawl on water quality in eastern Massachusetts, USA. *Environ Manag* 2007;**40**(2):183–200.
100. Delfanti L, Colantoni A, Recanatesi F, Bencardino M, Sateriano A, Zambon I, et al. Solar plants, environmental degradation and local socioeconomic contexts: a case study in a Mediterranean country. *Environ Impact Assess Rev* 2016;**61**:88–93.
101. Lanfredi M, Coppola R, D'Emilio M, Imbrenda V, Macchiato M, Simoniello T. A geostatistics-assisted approach to the deterministic approximation of climate data. *Environ Model Softw* 2015;**66**:69–77. 589. https://doi.org/10.1016/j.envsoft.2014.12.009.

102. Kosmas C, Karamesouti M, Kounalaki K, Detsis V, Vassiliou P, Salvati L. Land degradation and long-term changes in agro-pastoral systems: An empirical analysis of ecological resilience in Asteroussia-Crete (Greece). *Catena* 2016;**147**:196–204.
103. Angel S, Parent J, Civco DL, Blei A, Potere D. The dimensions of global urban expansion: estimates and projections for all countries, 2000–2050. *Prog Plan* 2011;**75**(2):53–107.
104. Balchin P. *Housing policy in Europe*. London and New York: Routledge; 1996.
105. Brouwer FB, Thomas AJ, Chadwick MJ. *Land use changes in Europe. Processes of change, environmental transformations and future patterns*. Dordrecht: Kluwer Academic Publishers; 1991.
106. Burchell RW, Downs A, McCann B, Mukherji S. *Sprawl costs-economic impacts of unchecked development*. Washington: Island Press; 2005.
107. Camagni R, Capello R, Nijkamp P. Toward sustainable city policy: an economy-environment-technology nexus. *Ecol Econ* 1998;**24**:103–18.
108. Couch C, Leontidou L, Petschel-Held G. *Urban sprawl in Europe—landscapes, land-use change & policy*. Oxford, UK: Blackwell Publishing; 2007.
109. Li X, Gong P. Urban growth models: progress and perspective. *Sci Bull* 2016;**61**(21):1637–50.
110. Claus T, Farinós-Dasí J, Llausàs A, Zornoza-Gallego C, Evers D. *SUPER—sustainable urbanisation and land use practices in european regions annex 3.3: case study BE-flanders*; 2020.
111. EEA (European Environmental Agency). *Land take and land degradation in functional urban areas—EEA report no. 17/2021*. Luxembourg: Publications Office of the European Union; 2022. ISBN:978-92-9480-465-5. available at https://www.iflaeurope.eu/assets/docs/TH-AL-22-001-EN-N_Land_take_and_land_degradation.pdf.
112. Bart IL. Urban sprawl and climate change: a statistical exploration of cause and effect, with policy options for the EU. *Land Use Policy* 2010;**27**(2):283–92.
113. Thirsk J, Whetham EH. *The agrarian history of England and Wales*. vol. 8. Cambridge University Press; 1978. p. 1914–39.
114. Lesthaeghe R. A century of demographic change in Western Europe: an exploration of underlying dimensions. *Popul Dev Rev* 1983;**9**:411–36.
115. Primdahl J, Andersen E, Swaffield S, Kristensen L. Intersecting dynamics of agricultural structural change and urbanisation within European rural landscapes: change patterns and policy implications. *Landsc Res* 2013;**38**(6):799–817.
116. Bontje M. Dealing with deconcentration: population deconcentration and planning response in polynucleated urban regions in north-west Europe. *Urban Stud* 2001;**38**(4):769–85.
117. Morys M, editor. *The economic history of Central, East and South-East Europe: 1800 to the present*. London: Routledge; 2020.
118. Graham E. Theory and explanation in demography: the case of low fertility in Europe. *Popul Stud* 2021;**75**(suppl. 1):133–55.
119. Favell A. The new face of East–West migration in Europe. *J Ethn Migr Stud* 2008;**34**(5):701–16.
120. Tosics I, Hegedüs J. Housing in south-eastern Europe. In: *Housing change in East and Central Europe*. Routledge; 2017. p. 21–44.
121. Reurink A, Garcia-Bernardo J. Competing for capitals: the great fragmentation of the firm and varieties of FDI attraction profiles in the European Union. *Rev Int Polit Econ* 2021;**28**(5):1274–307.
122. Mather AS. The forest transition. *Area* 1992;**24**:367–79.
123. Alonso-Sarría F, Martínez-Hernández C, Romero-Díaz A, Cánovas-García F, Gomariz-Castillo F. Main environmental features leading to recent land abandonment in Murcia region (Southeast Spain). *Land Degrad Dev* 2016;**27**(3):654–70.
124. Perpiña Castillo C, Coll Aliaga E, Lavalle C, Martínez Llario JC. An assessment and spatial modelling of agricultural land abandonment in Spain (2015–2030). *Sustainability* 2020;**12**(2):560.

125. Quaranta G, Salvia R, Salvati L, Paola VD, Coluzzi R, Imbrenda V, et al. Long-term impacts of grazing management on land degradation in a rural community of Southern Italy: Depopulation matters. *Land Degrad Dev* 2020;**31**(16):2379–94.
126. Simoniello T, Coluzzi R, Imbrenda V, Lanfredi M. Land cover changes and forest landscape evolution (1985–2009) in a typical Mediterranean agroforestry system (high Agri Valley). *Nat Hazards Earth Syst Sci* 2015;**15**(6):1201–14.
127. Smith P, Gregory PJ, Van Vuuren D, Obersteiner M, Havlík P, Rounsevell M, et al. Competition for land. *Philos Tran R Soc B Biol Sci* 2010;**365**(1554):2941–57.
128. Zhongming Z, Linong L, Xiaona Y, Wangqiang Z, Wei L. *Land in Europe: prices, taxes and use patterns, European Environment Agency Kongens Nytorv 6, 1050.* Denmark: Copenhagen K; 2010.
129. Wigier M. The competitiveness of Polish agriculture after accession to the EU. *Економика пољопривреде* 2014;**61**(1):87–102.
130. Atu JE, Ayama OR, Eja EI. Urban sprawl effects on biodiversity in peripheral agricultural lands in Calabar, Nigeria. *J Environ Earth Sci* 2013;**3**(7):219–31.
131. Zhou Y, Xing B, Ju W. Assessing the impact of urban sprawl on net primary productivity of terrestrial ecosystems using a process-based model—a case study in Nanjing, China. *IEEE J Select Top Appl Earth Observ Remote Sens* 2015;**8**(5):2318–31.
132. Bajocco S, Ceccarelli T, Smiraglia D, Salvati L, Ricotta C. Modeling the ecological niche of long-term land use changes: the role of biophysical factors. *Ecol Indic* 2016; **60**:231–6.
133. Salvati L, Zitti M, Sateriano A. Changes in city vertical profile as an indicator of sprawl: evidence from a Mediterranean urban region. *Habitat Int* 2013;**38**:119–25.
134. Salvati L, Petitta M, Ceccarelli T, Perini L, Di Battista F, Scarascia MEV. Italy's renewable water resources as estimated on the basis of the monthly water balance. Irrigation and drainage. *J Int Commission Irrig Drain* 2008;**57**(5):507–15.
135. Bogunovic I, Pereira P, Kisic I, Sajko K, Sraka M. Tillage management impacts on soil compaction, erosion and crop yield in Stagnosols (Croatia). *Catena* 2018;**160**:376–84.
136. Coluzzi R, Bianchini L, Egidi G, Cudlin P, Imbrenda V, Salvati L, et al. Density matters? Settlement expansion and land degradation in peri-urban and rural districts of Italy. *Environ Impact Assess Rev* 2022;**92**:106703. https://doi.org/10.1016/j.eiar.2021.106703.
137. Halbac-Cotoara-Zamfir R, Smiraglia D, Quaranta G, Salvia R, Salvati L, Giménez-Morera A. Land degradation and mitigation policies in the Mediterranean region: a brief commentary. *Sustainability* 2020;**12**(20):8313.
138. Imbrenda V, D'Emilio M, Lanfredi M, Ragosta M, Simoniello T. Indicators of land degradation vulnerability due to anthropic factors: tools for an efficient planning. In: Borruso G, Bertazzon S, Favretto A, Murgante B, Torre CM, editors. *Geographic information analysis for sustainable development and economic planning: new technologies*. Hershey: IGI Global; 2013. p. 87–101. https://doi.org/10.4018/978-1-4666-4852-4.ch078.
139. Imbrenda V, Coluzzi R, Di Stefano V, Egidi G, Salvati L, Samela C, et al. Modeling spatio-temporal divergence in land vulnerability to desertification with local regressions. *Sustainability* 2022;**14**(17):10906.
140. Kairis O, Karavitis C, Salvati L, Kounalaki A, Kosmas K. Exploring the impact of overgrazing on soil erosion and land degradation in a dry Mediterranean agro-forest landscape (Crete, Greece). *Arid Land Res Manag* 2015;**29**(3):360–74.
141. Nickayin SS, Coluzzi R, Marucci A, Bianchini L, Salvati L, Cudlin P, et al. Desertification risk fuels spatial polarization in 'affected' and 'unaffected' landscapes in Italy. *Sci Rep* 2022;**12**:747. https://doi.org/10.1038/s41598-021-04638-1.
142. Seifollahi-Aghmiuni S, Kalantari Z, Egidi G, Gaburova L, Salvati L. Urbanisation-driven land degradation and socioeconomic challenges in peri-urban areas: insights from Southern Europe. *Ambio* 2022;**51**(6):1446–58.
143. Salvati L, Gemmiti R, Perini L. Land degradation in Mediterranean urban areas: an unexplored link with planning? *Area* 2012;**44**(3):317–25.

144. Ienciu I, Vorovencii I, Oprea L, Popescu C. Urban development of mountain areas with the aim of developing local tourism. *J Environ Prot Ecol* 2013;**14**(3):980–5.
145. Tulla AF, Pallarès-Barberà M, Vera A. Naturbanization and local development in the mountain areas of the Catalan Pyrenees. In: *Naturbanization*. CRC Press; 2008. p. 83–100.
146. Bender O, Kanitscheider S. New immigration into the European Alps: emerging research issues. *Mt Res Dev* 2012;**32**(2):235–41.
147. Boori MS, Voženílek V, Choudhary K. Land use/cover disturbance due to tourism in Jeseníky Mountain, Czech Republic: a remote sensing and GIS based approach. *Egypt J Remote Sens Space Sci* 2015;**18**(1):17–26.
148. Newman P, Kenworthy J. *Sustainability and cities: overcoming automobile dependence*. Washington, DC: Island Press; 1999.
149. Ciscel DH. The economics of urban sprawl: inefficiency as a core feature of metropolitan growth. *J Econ Issues* 2001;**35**(2):405–13.
150. Andong RF, Sajor E. Urban sprawl, public transport, and increasing CO2 emissions: the case of Metro Manila, Philippines. *Environ Dev Sustain* 2017;**19**(1):99–123.
151. Gielen E, Riutort-Mayol G, Miralles i Garcia JL, Palencia Jimenez JS. Cost assessment of urban sprawl on municipal services using hierarchical regression. *Environ Plan B Urban Anal City Sci* 2021;**48**(2):280–97.
152. Beck HE, Zimmermann NE, McVicar TR, Vergopolan N, Berg A, Wood EF. Present and future Köppen-1013 Geiger climate classification maps at 1-km resolution. *Sci Data* 2018;**5**:180214. https://doi.org/10.1038/sdata.2018.214.1014.
153. Lanfredi M, Coluzzi R, Imbrenda V, Macchiato M, Simoniello T. Analyzing space–time coherence in 1015 precipitation seasonality across different European climates. *Remote Sens* 2020;**12**:171. https://doi.org/10.3390/rs12010171.
154. Coluzzi R, Fascetti S, Imbrenda V, Italiano SSP, Ripullone F, Lanfredi M. Exploring the use of sentinel-2 1071 data to monitor heterogeneous effects of contextual drought and heatwaves on Mediterranean forests. *Land* 2020;**9**:325. https://doi.org/10.3390/land9090325.
155. Greco S, Infusino M, De Donato C, Coluzzi R, Imbrenda V, Lanfredi M, et al. Late spring frost in Mediterranean beech forests: extended crown dieback and short-term effects on moth communities. *Forests* 2018;**9**:388.
156. Manfreda S, Samela C, Gioia A, Consoli GG, Iacobellis V, Giuzio L, et al. Flood-prone 1023 areas assessment using linear binary classifiers based on flood maps obtained from 1D and 2D hydraulic 1024 models. *Nat Hazards* 2015;**79**:735–54. https://doi.org/10.1007/s11069-015-1869-5.
157. Faccini F, Luino F, Paliaga G, Sacchini A, Turconi L, de Jong C. Role of rainfall intensity and urban sprawl in the 2014 flash flood in Genoa City, Bisagno catchment (Liguria, Italy). *Appl Geogr* 2018;**98**:224–41.
158. Djordjević S, Butler D, Gourbesville P, Mark O, Pasche E. New policies to deal with climate change and other drivers impacting on resilience to flooding in urban areas: the CORFU approach. *Environ Sci Pol* 2011;**14**(7):864–73.
159. Ciscar JC, Iglesias A, Feyen L, Szabó L, Van Regemorter D, Amelung B, et al. Physical and economic consequences of climate change in Europe. *Proc Natl Acad Sci* 2011;**108**(7):2678–83.
160. Jarratt D, Davies NJ. Planning for climate change impacts: coastal tourism destination resilience policies. *Tour Plan Devel* 2020;**17**(4):423–40.
161. De Bruijn KM, Klijn F. Risky places in the Netherlands: a first approximation for floods. *J Flood Risk Manage* 2009;**2**(1):58–67.
162. Kellens W, Vanneuville W, Verfaillie E, Meire E, Deckers P, De Maeyer P. Flood risk management in Flanders: past developments and future challenges. *Water Resour Manag* 2013;**27**(10):3585–606.
163. Paprotny D, Terefenko P. New estimates of potential impacts of sea level rise and coastal floods in Poland. *Nat Hazards* 2017;**85**(2):1249–77.

164. Kulp SA, Strauss BH. New elevation data triple estimates of global vulnerability to sea-level rise and coastal flooding. *Nat Commun* 2019;**10**(1):1–12.
165. Neumann B, Vafeidis AT, Zimmermann J, Nicholls RJ. Future coastal population growth and exposure to sea-level rise and coastal flooding-a global assessment. *PLoS One* 2015;**10**(3):e0118571.
166. Gallagher P. The environmental, social, and cultural impacts of sprawl. *Nat Resour Environ* 2001;**15**(4):219–67.
167. Oueslati W, Alvanides S, Garrod G. Determinants of urban sprawl in European cities. *Urban Stud* 2015;**52**(9):1594–614.
168. Elizondo GM, Lofthouse V. Towards a sustainable use of water at home: understanding how much, where and why. *J Sustain Devel* 2010;**3**(1):3–10.
169. Haase D, Kabisch N, Haase A. Endless urban growth? On the mismatch of population, household and urban land area growth and its effects on the urban debate. *PLoS One* 2013;**8**(6):e66531.
170. Sharp Jr JM, Krothe JN, Mather JD, Gracia-Fresca B, Stewart CA. Effects of urbanization on groundwater systems. In: *Earth science in the city: a reader*. vol. 56. Wiley Online Library; 2003. p. 257–78.
171. Mihai B, Nistor C, Simion G. Post-socialist urban growth of Bucharest, Romania–a change detection analysis on Landsat imagery (1984–2010). *Acta Geogr Slov* 2015;**55**(2):223–34.
172. Grădinaru SR, Iojă CI, Onose DA, Gavrilidis AA, Pătru-Stupariu I, Kienast F, et al. Land abandonment as a precursor of built-up development at the sprawling periphery of former socialist cities. *Ecol Indic* 2015;**57**:305–13.
173. Oliver L, Ferber U, Grimski D, Millar K, Nathanail P. The scale and nature of European brownfields. In: *CABERNET 2005—international conference on managing urban land LQM Ltd., Nottingham, Belfast, Northern Ireland, UK*; 2005, April. p. 5–6.
174. Camagni R, Gibelli MC, Rigamonti P. Urban mobility and urban form: the social and environmental costs of different patterns of urban expansion. *Ecol Econ* 2002;**40**(2):199–216.
175. Lamonica GR, Recchioni MC, Chelli FM, Salvati L. The efficiency of the cross-entropy method when estimating the technical coefficients of input–output tables. *Spat Econ Anal* 2020;**15**(1):62–91.
176. Rodríguez MC, Dupont-Courtade L, Oueslati W. Air pollution and urban structure linkages: evidence from European cities. *Renew Sust Energ Rev* 2016;**53**:1–9.
177. EEA (European Environmental Agency). *Air quality in Europe 2020 report, Kongens Nytorv 6 1050 Copenhagen K, Denmark*; 2020. available at https://www.eea.europa.eu/publications/air-quality-in-europe-2020-report.

Further reading

178. Scott AJ. *Global city-regions—trends, theory, policy*. New York: Oxford University Press; 2001.
179. United Nations. *World urbanization prospects: the 2006 revision*. New York: UN; 2007.

CHAPTER THREE

Hydrological challenges in urban areas

Carla S.S. Ferreira[a,b,c,*], António C. Duarte[d], Milica Kasanin-Grubin[e], Marijana Kapovic-Solomun[f], and Zahra Kalantari[a,b,g]

[a]Department of Physical Geography and Bolin Centre for Climate Research, Stockholm University, Stockholm, Sweden
[b]Navarino Environmental Observatory, Messenia, Greece
[c]Research Center of Natural Resources, Environment and Society, Polytechnic Institute of Coimbra, Coimbra School of Agriculture, Coimbra, Portugal
[d]School of Agriculture/Polytechnic Institute of Castelo Branco, Branco, Portugal
[e]Institute of Chemistry, Technology and Metallurgy, University of Belgrade, Belgrade, Serbia
[f]University of Banja Luka, Faculty of Forestry, Department of Forest Ecology, Banja Luka, Republic of Srpska, Bosnia & Herzegovina
[g]Department of Sustainable Development, Environmental Science and Engineering (SEED), KTH Royal Institute of Technology, Stockholm, Sweden
*Corresponding author: e-mail address: carla.ferreira@natgeo.su.se

Contents

Abstract

Urbanization alters hydrological processes and is often associated with increasing flood risks, which threaten human wellbeing and social and economic development. The conventional paradigm of flood protection relying on structural measures based on hard engineering solutions (e.g., dams, piped systems) has proven insufficient to mitigate floods. Sustainable water management, including solutions to enhance natural processes within urban areas, is a promising approach to enhance flood resilience and address the multiple sustainability challenges faced by cities. However, implementation of solutions based on mimicking natural processes has been slow. Mainstreaming of urban sustainable flood management is inhibited by governance aspects (e.g., limited collaborative governance), and knowledge gaps on effectiveness compared with

Advances in Chemical Pollution, Environmental Management and Protection, Volume 8
ISSN 2468-9289
https://doi.org/10.1016/bs.apmp.2022.09.001

conventional engineering approaches. The increasing flood hazards driven by growing urban populations and climate change projections of increasing frequency and intensity of large precipitation events demand improvements in spatial planning. This also provides opportunities for sustainable water management mainstreaming in order to complement the relatively limited drainage capacity of conventional systems.

Keywords: Urban areas, Hydrological processes, Flood risk assessment, Sustainable water management, Flood resilience

1. Introduction

The global population is currently approximately 7.8 billion and around 56.2% live in urban areas.[1] In Europe, the population living in urban areas exceeds 80% and is still displaying a growing trend, which is expected to stabilize at around 90% by 2030.[2] The majority of urban residents in Europe are concentrated in cities with a population of between 250,000 and 5,000,000.[3] European cities are dense compared with those in other parts of the world, with an average of 3000 residents per km^2 in the main medium-size cities.[3] During the 19th and 20th century, most of the world's urban population and of its largest cities were in prosperous nations, but nowadays low- and middle-income nations are home to close to two-fifths of the world's total population, close to three-quarters of its urban population, and most of its large cities.[4] Although population growth is generally declining, urban areas are still growing and the rural population is decreasing, with an annual loss of 1.5%.[5] The increase in the world's urban population and in large cities outside high-income nations represent important current changes. By 2050, 70% of the world's population will live in cities, making developments in urbanizing regions critical in achieving a sustainable future for the world.[6]

Increasing human pressure in urban areas poses a number of environmental and social challenges.[7] At the present time, cities are the subject of planning debates because of their rapid growth and associated changes in landscapes, their impacts in terms of natural and climate-related hazards, environmental degradation (e.g., increasing levels of air and water pollution and waste production), and increasing social inequities.[8] Intensive urbanization, which is associated with extensive soil sealing, and climate change are the most significant drivers of changes in the water cycle, which alter the hydrological regime and increase the flood risk.[9,10] Cities typically at the highest risk from extreme weather events such as floods are mainly those in high-income countries.[11] Floods are one of the most frequent natural disasters, and are responsible for 1–1000 fatalities per event.[12] In Europe, as in other world regions, the highest average mortality per event is caused by flash floods, which, e.g., between 1975 and 2001 led to average mortality of 5.6%, while the average mortality rate for river floods was 0.47%.[12] Although flash floods generally affect a limited number of people compared with other types of flood they tend to be deadlier, so development of good warning and prediction systems is important.[12]

Water management is becoming an issue of particular importance in urban areas.[13] Unfortunately, at many sites of rapid urbanization, water resource management takes place at a later phase (e.g., during recovery after a flood event). Thus, sustainable water management planning relies on inappropriate coordination and local management as the city expands.[14] Urbanization, particularly on flood plains, has a direct impact on flood risk because it affects the invested values and may also increase water levels during floods.[15] This increases the importance of spatial planning policy.[16] Furthermore, urban areas must carefully manage their water, since quantity and quality problems are both expected to increase under future climate change.[17]

The need for strategies for prevention and mitigation of floods and runoff in urban areas is increasing interest in sustainable urban water management and in expanding the capacity of water infrastructure facilities in parallel with the urbanization process. Existing adaptation and mitigation measures by cities tend to focus on the performance of the city's infrastructure and institutional systems, including communication, and integration of multiple sectors, and risk management in a dynamic physical, economic, institutional, and socio-political environment.[18] Presence and good efficiency of early warning systems, capability to react, and high effectiveness of post-disaster response are especially important for urban areas identified as more sensitive and with less coping capacity. On the other hand, improved resilience of urban areas requires synchronized actions by institutions or higher levels of government, e.g., watershed management upstream of a city to reduce flood risks.[19,20] Urban governments play an important role in urban resilience to climate change, although their ability to act usually depends on legislative and financial support from higher levels of government. Cities in high-income nations and many in middle-income nations have become more resilient to extreme weather through a range of measures responding to risks and political processes demanding such responses.[2] On the other hand, many cities with a good level of resilience may still not be equipped to respond to the future changes in hazards and risks associated with climate change.[17] Thus, the European Floods Directive (2007/60/EC) raises the need for a shift from a conventional flood protection approach to a risk-based approach, taking different risk levels into account and designing measures based on the potential damage.

This chapter describes the impacts of urbanization on hydrological processes and flood hazards, and discusses different stormwater management strategies that can be used to mitigate floods in urban areas. This is followed by a deliberation on the challenges posed by global-scale changes (i.e., urbanization and climate change) to current urban water management approaches and the need to mitigate and adapt for flood risk in the future.

2. Hydrological processes in urban areas

2.1 Catchment characteristics determining hydrological processes

The hydrological cycle is generally defined as a conceptual model describing the storage and circulation of water between the biosphere, atmosphere, lithosphere, and the hydrosphere.[21] In open green areas, the soil water balance can be formulated as follows[22]:

$$RO_i = (RZWD_i - RZWD_{i-1}) + R_i - ET_i - DP_i \quad (1)$$

where RO_i is runoff (mm), $RZWD_i$ and $RZWD_{i-1}$ are root zone soil water depletion on days i and i-1, respectively, and R_i, ET_i, and DP_i (all in mm) are rainfall, crop evapotranspiration, and deep percolation, respectively, on day i.

One major consequence of intensive urbanization is the expansion in sealed/paved surfaces, e.g., roads, buildings, and parking lots. These changes affect landscape dynamics and in particular water fluxes, leading to the disruption of natural hydrological processes. In urbanized areas, the hydrological cycle becomes more complex due to many anthropogenic influences and interventions, and is normally called the urban water cycle. Relative to the water cycle in natural areas, the most significant changes in physical aspects of hydrology in urban areas include: (i) removal of natural vegetation (trees, shrubs, herbaceous vegetation) in open areas and along the banks of streams and rivers; (ii) removal of local depressions that can temporarily store surface water; (iii) decreased soil infiltration capacity and consequently loss of water retention; (iv) increasing impervious areas (e.g., residential, industry, roads, airports); and (v) construction of artificial drainage systems (i.e., sewer network, pluvial network, open channels, culverted streams).

Thus, even though the components of the hydrological cycle are the same, their effects in urban environments are quite different in the following regards: (i) reduced effects of vegetation in rainfall interception and evapotranspiration[23]; (ii) decreased water infiltration into the soil and thus reduced recharge of subsurface runoff and aquifers; (iii) decreased surface roughness, leading to increasing runoff volume and velocity, associated with increasing runoff energy and power for transport capacity and soil erosion[24]; and (iv) increasing peak flows and frequency of flash floods, with shorter time of concentration and reduced recession times[25] (Fig. 1). These changes in the streamflow regime impact urban aquatic ecosystems, for example by reducing the physical size and productivity of the ecosystems.[26]

Typically, surface runoff increases with the percentage of impervious surface in a city. At the same time, deep percolation decreases relative to soil natural coverage, also in line with the extent of impervious surface, but to a higher percentage than the infiltration decreases. The decrease in evapotranspiration is not as large as those in infiltration and deep percolation, even with the loss of vegetation, because evaporation from sealed surfaces remains relatively high in the urban environment. The high rates of energy fluxes in cities, e.g., gaseous emissions from traffic, contribute to the urban heat island effect, characterized by higher air temperatures, and thus evaporation rates, in urban areas.[27]

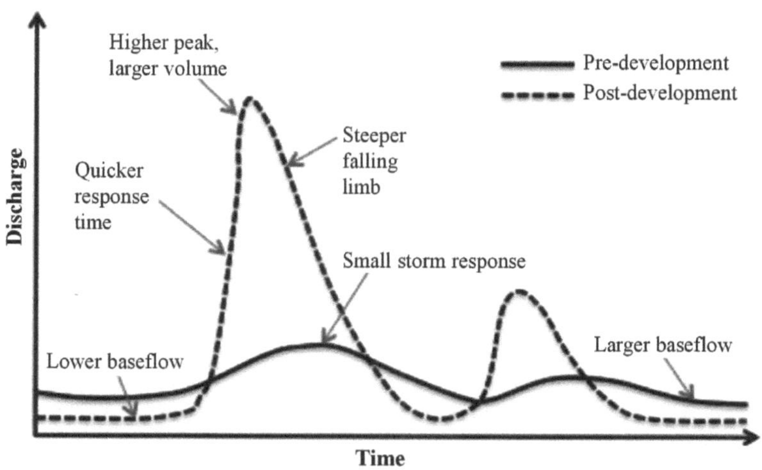

Fig. 1 Schematic illustration of the impacts of urbanization on hydrograph shape. *Adapted from Fletcher TD, Andrieu H, Hamel, P. Understanding, management and modelling of urban hydrology and its consequences for receiving waters: a state of the art.* Adv Water Resour *2013;**51**:261–79.*

2.2 The role of flow connectivity in hydrological responses

There is general consensus in the literature on the negative impacts of urbanization on the hydrological cycle, particularly on streamflow changes (Fig. 1), but the magnitude of the impacts varies between studies. For example, in a study in a USA catchment, an increase in urban surface from 20% to 100% led to a 50% increase in total runoff and a 50% reduction in actual evapotranspiration and percolation.[28] In southern Alberta, Canada, a 65% increase in built-up area caused decreases of only 1% and 2.3% in total evapotranspiration and water infiltration, respectively. These changes led to a 7.3% increase in surface flow, but also a 13.2% decrease in baseflow, resulting in a total flow decrease of 4%.[29] In Western Australia, 20% urbanization of a natural area led to an infiltration decrease from 80% to less than 20%, causing a decline in groundwater level and an increase in runoff coefficient from 1% to more than 40%.[30] In the Big River catchment in east-central Missouri, a three-fold increase in urban area over a 15-year period resulted in a 140% increase in high flow events and a decrease in frequency of low flow events of up to 100%.[31]

The differing magnitude of the impacts of urbanization at different sites may be a consequence of site-specific (i) spatio-temporal patterns of runoff processes generated within the catchments, and (ii) flow connectivity between sources and the stream network. In urban areas, runoff generated on impervious surfaces may be totally or partially retained and/or infiltrated in pervious surfaces.[32] Additionally, runoff occurs not only due to intensive precipitation on sealed surfaces but can also be generated on pervious surfaces, due to overland flow when precipitation intensity exceeds soil infiltration capacity. The point at which this occurs depends on soil properties, such as unsaturated hydraulic conductivity, land-use history and management activities, and rainfall pattern.[33] Urban soils are usually associated with lower infiltration capacity, due to physical degradation through compaction.[34] Moreover, runoff sources in permeable areas are affected by topography and antecedent soil moisture content.[32] As noted by Duarte et al.,[35] high antecedent soil moisture conditions significantly decrease the infiltration rate and limit soil capacity for water retention, and thus have a great influence on the magnitude of runoff events. During wet periods, overland flow can be also generated in saturated areas of permeable soils.[32]

Within urban areas, if overland flow from impervious surfaces flows onto pervious surfaces, it may infiltrate before reaching the catchment drainage network.[36] Thus, streamflow response will depend on the extent and distribution of impervious and pervious surfaces, and on the connectivity between land surface and the drainage network, driven by the spatial form and location of different land uses.[32,37] Slopes can therefore behave as a mosaic of runoff and run-on areas, providing non-uniform infiltration. On each surface, interception and depression storage must be satisfied before overland flow commences. Initial losses are known to be small on impervious surfaces,[38] but larger on pervious areas.[36] As a consequence, overland flow from pervious sites is more difficult to predict than overland flow from impervious surfaces. These factors influence the landscape structure and spatial organization of a catchment which, in turn, determine the distribution of water flow paths, the patterns of water storage, and residence time distribution.[39] Considering the relevance of the extent and location of pervious and impervious surfaces to overland flow and runoff generation, understanding and quantification of the hydrological impacts of urbanization require a detailed characterization of different land cover types.[40]

The distance between overland flow sources (pervious soils and/or impervious surfaces) and the drainage network (main channel or tributaries) is an important parameter influencing streamflow response (e.g.,[41]). Source areas located close to the drainage network can be significant contributors to runoff, while those located farther away may have little or no impact on streamflow, due to greater opportunities for surface flow retention and infiltration over the hillslope.[42] The lack of flow connectivity between runoff sources and the stream network has been used to explain unexpected relationships between total impervious area (TIA) and streamflow parameters.[37] In order to overcome these problems, some hydrological studies have included an additional parameter, effective impervious area (EIA), which represents the impervious areas directly connected to the stream network.[34]

Overland flow from directly connected impervious surfaces will reach the slope outlet more rapidly than runoff from impervious surfaces onto areas with significant capacity for abstraction or storage. If the runoff from upslope impervious surfaces is not directly connected with the outlet, it will only contribute to streamflow if downslope soil infiltration capacity or water storage capacity are exceeded by rainfall and generate run-on. During rainfall events, as the downslope soil infiltration capacity and/or storage capacities decline, a pervious soil surface may generate overland flow and can become similar to an impervious surface.[43]

Besides the spatial distribution of pervious and impervious surfaces within a catchment, flow connectivity is also affected by water management activities.[44] Problems with urban runoff are usually managed with engineered solutions associated with channelization of water. In urban/peri-urban areas there are three basic types of drainage systems: (1) sanitary sewerage for domestic and industrial wastewater, (2) storm drains intended to convey storm runoff rapidly and safely, and (3) combined sewerage, which drains wastewater and storm runoff in one system. The introduction of artificial drainage increases the direct input of precipitation to stream channels, by circumventing depression storage and groundwater recharge.[45] Storm runoff channelization leads to a faster rise and recession of streamflow, higher peak rates, and increased storm flow volume from a given amount of precipitation.[46] The faster runoff resulting from storm drainage systems can induce flood risks in downstream areas, particularly in small catchments.[47] The complex interactions between factors affecting flow connectivity in the landscape (Fig. 2) and the hydrological response of urban catchments creates a demand for additional scientific information when examining the ways in which flow dynamics are changed by human impacts.

Hydrology is an important issue in the urban environment and the capacity to predict critical rainfall events for runoff generation is particularly critical. Normally, event prediction is addressed through rainfall evaluation, forecasting, and modeling. Many innovative tools have emerged for recording, analyzing, and predicting rainfall, such as radar technologies (radar images) that give information on rainfield displacement and meteorological models that consider rainfall dynamics, with applications in rainfall-runoff models.[48,49] Rainfall-runoff models can vary in temporal scale (time step, event-specific or continuous) and spatial scale (lumped or spatially

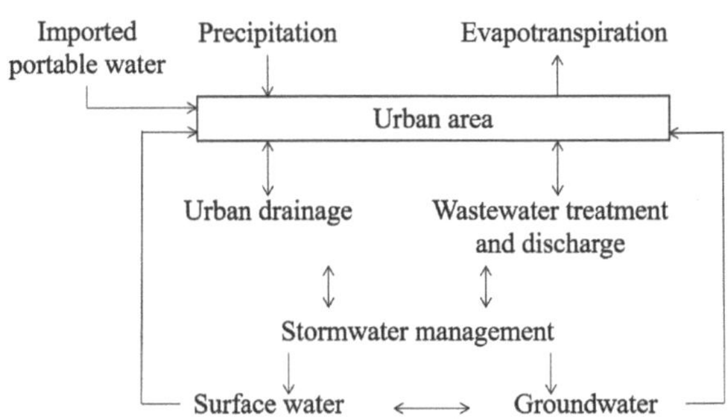

Fig. 2 The urban water cycle and its main components and pathways. *Adapted from Marsalek JB, Jiménez-Cisneros E, Malmquist PA, Karamouz M, Goldenfum J, Chocat B. Urban water cycle processes and interactions.* IHP-VI technical documents in hydrology no. 78, *Paris: UNESCO; 2006.*

distributed).[50] At present, simulation of urban hydrology relies on spatially distributed approaches. These have been greatly enhanced by the information collected by remote sensing and spatial databases, which have allowed more accurate definition of land use areas and other landscape characteristics that facilitate the simulation process.[37] According to Ferrans et al.,[20] the Storm Water Management Model (SWMM)[51] is one of the most frequently used, followed by the LTHIA-LID model.[52] The most basic approaches used in existing analytical models consist of water balance calculations, statistical analyses, or GIS-based tools.

Very important issues when assessing the accuracy and robustness of model results are sensitivity analysis, model calibration, and model validation.[53] Another important issue is recognition of model limitations, namely the uncertainty that it brings (e.g., values attributed to model parameters), data input and calibration data, and the most appropriate mode in which to run the model in specific situations.[54] Even with recent advances in model design, prediction of future urban hydrology remains an important challenge, particularly given the uncertainties associated with climate change.

3. Flood management approaches

3.1 Flood risk assessment and flood management

Urban water management is one of humanity's great challenges,[55] particularly regarding stormwater, since flood events appear to be occurring more frequently.[56] Flooding is becoming more important particularly in developing countries, given the complex socio-economic factors and the interactions between multiple stakeholders affecting the dynamics of flood risk.[56] Flood risk assessment involves analyzing and understanding flood hazard, which represents the probability that a flood will occur over a time horizon, under existing conditions of vulnerability (those determined by physical, social, economic, and environmental factors or processes which increase the susceptibility to flooding) and exposure (i.e., the situation of human assets located in hazard-prone areas), and cause harm to urban dynamics (e.g., impacts on human health, economic activities, and environmental impacts).[57,58] Identification of hazard-prone areas typically involves estimation of hydrological variables (e.g., water depth, flow velocity) and mapping, often through modeling. The ability for accurate determination is dependent on available data to run the models, i.e., data on areas affected by past floods (e.g., based on past records and/or remote sensing). Poor data availability can be a major problem in flood hazard assessment, particularly in developing countries where e.g., weather stations are often lacking or inadequate, resulting in scarcity of data on the actual frequencies, intensities, and impacts of flood events.[56]

People living in poor urban communities, especially those in developing countries, such as Senegal and Lebanon, are often more exposed to urban flooding, since they tend to be located in the highest flood hazard areas (e.g., along banks of rivers, tidal streams, and drainage canals), usually in informal settlements with high population density and consisting of households that are less resilient to flood events.[56] Coastal settlements are also highly exposed to flood risk, such as those located in the Mediterranean region[59] and in the Pearl River Delta region of China, where more than 42.4 million people are exposed to flood risk and less than 10% of coastal areas are protected against flood events with a 1- to 50-year return period.[57]

Flood risk assessment requires reliable data on the nature, trends, and spatial distribution of flood hazards, technology to predict flood occurrence, and structured early warning and evacuation systems.[56] Methods for flood risk assessment generally consist of (i) methods based on existing statistics on historical disasters, usually made available by governments and authorities but possibly not consistent with reality, e.g., they may not represent appropriately the rapid changes occurring in urban areas[60]; (ii) scenario simulation analysis, typically reliant on two-dimensional hydraulic/hydrodynamic models to provide inundation outputs in terms of floodwater depth and velocity[61,62]; (iii) multi-criteria decision analysis, based e.g., on scoring by experts (using their experience and knowledge, resulting in subjectivity and uncertainty in the assessment results), set pair analysis, and the analytic hierarchy process[63]; and (iv) machine learning models, driven by increasing computational advances and algorithms.[60] Particularly in informal settlements in developing countries, lack of data on the current population, housing, and infrastructure is often a major constraint for flood risk assessment.[56]

Based on flood risk assessment, appropriate measures for mitigating risk can be designed. These may include strategies for application in pre-disaster, during, and post-disaster phases, considering the particular socioeconomic conditions of local communities and the capabilities of the operating institutions.[64] Flood risk management in the pre-disaster phase is based on structural and non-structural measures, but generally involves defense-oriented techniques. Structural measures rely on hard-engineering control measures for flood defenses, such as storm drains and channels, to ensure rapid transport of water from the urban area, and large infrastructures such as dams and dykes to enhance water storage capacity.[56] Another type of measure involves subterraneous artificial drainage systems that can combine or separate the pluvial water system and the wastewater system, in a single or double pipe network, respectively. During heavy rainy events, combined drainage systems may experience overloading that exceeds their capacity, with the excess water draining to surrounding areas, leading to environmental pollution and degradation of aquatic ecosystems.[65]

Non-structural measures are based on land use planning to control and guide urban development away from potential risks.[64] Regulating the use of state and local planning requirements such as land zoning (e.g., preventing urban expansion into flood prone areas) and development standards and controls can be highly relevant for flood risk management.[66] In the UK, for example, policies to restrict new developments in areas at risk of flooding have been successfully implemented by giving local planning agencies the authority to reject applications for new developments if the proposal fails the flood risk appraisal, or to approve applications if the developer assumes responsibility for flood risk mitigation and ensures that the new development is flood-protected to acceptable standards.[67] Land use planning is considered a crucial component in managing flood risks.[68] Strategic spatial planning as a defense-orientated approach is challenging, however, given the temporal mismatch between long-term flood management goals and other planning considerations[68] and the difficulties in balancing public interest in flood water storage on private land and private property rights.[69]

Development design is also a relevant aspect of flood mitigation. The arrangement of impervious surfaces, a major component of urban planning, must enhance water infiltration and runoff storage capacity.[32] The layout of sealed surfaces alters flow paths and affects the magnitude of runoff and the response time in streamflow, as discussed in Section 2. Thus, it changes both the magnitude and spatial distribution of flood hazard.[33]

3.2 Sustainable flood management

Over recent years, there has been increasing recognition that, for effective mitigation of urban flood risk, it is necessary to go beyond conventional approaches relying on resistance-based strategies (i.e., flood defenses and urban drainage systems) and instead focus on flood-resilient urban planning.[70] Flood resilience covers the uncertainties associated with flood hazards by focusing on the ability of systems to absorb shocks.[71] Sustainable urban water management is an urban planning and design approach aiming to minimize the impacts of urbanization (see Section 2) by augmenting natural hydrological processes.[72] Several concepts, such as sustainable urban drainage solutions, green infrastructure, water-sensitive urban planning, sponge cities, and nature-based solutions, have been used in flood-resilient approaches.[70] These concepts employ a variety of measures and techniques to retain, detain, harvest, infiltrate, and filter runoff, in order to manage stormwater as close to its source as possible.[73] These near-natural stormwater management solutions also allow runoff to be delayed, and can thus increase evaporation via soil and plants, and support groundwater recharge.[65] Examples of such measures include constructed wetlands, riparian buffers, rain gardens, green roofs, and bioswales. In principle, they rely on integrating vegetated areas and open spaces within the urban network, and they can be applied in a variety of urban areas (e.g., residential, commercial), including new developments or redevelopments, and at a range of scales.[74]

Some scientists claim that solutions based on infiltration, delay, and/or storage of stormwater can be a full alternative to piped drainage, while others view them as additional measures to reduce pressure on conventional infrastructure (e.g.,[75]). However, there is increasing recognition that this type of approach focusing on natural processes must be combined with typical structural flood mitigation approaches.[76] Small-scale measures aimed at mimicking the natural hydrological cycle can complement large-scale measures in providing sufficient levels of flood protection in urban areas.[77] Besides flood mitigation, this type of measure provides a wide range of ecosystem services with multiple environmental, social, and economic benefits, such as enhanced biodiversity, esthetics, and human wellbeing.[78] However, measures based on natural processes require more space than conventional drainage solutions, which may be challenging in urban areas.[67]

Despite increasing recognition of the need to shift from conventional, large-scale centralized stormwater management infrastructures to distributed measures enabling sustainable stormwater management, implementation of sustainable approaches has been slow.[58] Several barriers have been identified for this slow implementation.[79] Urban water management involves a diverse range of political, economic, social, technological, legal, and environmental influences, and thus difficulties in coordinating cross-sector governance and collaboration between urban agencies is a key cause.[56] This is more relevant in countries with complex institutional structures for flood management, involving several institutions, agencies, and departments under different ministries.[80] Inter-sectoral and stakeholder collaboration is required to face challenges regarding, for example, space and cost constraints to planning, designing, and implementing sustainable urban water management systems.[66] Several sustainable water management projects have failed to operationalize fully due to pragmatic economic considerations, such as market mechanisms which protect water agency revenues, although greater economic value can be generated by avoiding floods and overloading conventional stormwater drainage systems.[81] Additional socio-economic research is required to better support new funding schemes for sustainable water management projects.[55]

Another barrier to sustainable stormwater management mainstreaming is limited knowledge on the effectiveness of sustainable stormwater measures compared with conventional alternatives.[82] Future implementation can be supported by improving knowledge of different hydrological, environmental, socio-economic, and management conditions, and by providing well-established historical data on positive impacts to create a stronger evidence base.[83] The current lack of knowledge makes it difficult to determine where to invest resources to improve urban water management projects. This knowledge barrier is particularly relevant in developing countries, where hydrology and urban flooding are poorly understood, and thus flood risk management measures are either lacking or not adequately put in place.[82] Filling the knowledge gap through targeted research can be useful in promoting private sector investment in sustainable water management systems, and in advocating policy changes supporting and promoting their implementation.[84]

4. Challenges raised by global changes

Increasing urbanization associated with the growing global population will lead to expansion of surface sealing within existing urban areas and beyond their periphery.[38] As explained in Section 2, these changes have implications for runoff processes associated with increasing flood hazard. However, one must bear in mind that the concentration of people in urban areas also provides various possibilities to reduce these impacts and transition towards sustainable development. Implementation of international agreements such as 2030 Agenda for Sustainable Development Goals, the Paris Agreement on Climate Change, and the New Urban Agenda is an opportunity to address risk drivers and focus on a number of problems brought about by urbanization.[85]

Climate change projections are clear about the future increases in both the intensity and frequency of precipitation,[86] which are expected to increase the flood risk substantially in coming years, particularly in central, northwest, and northern Europe.[87] Many European cities are following global recommendations on reducing the impact of climate change, and therefore national urban polices can be as a key instrument for implementation of international agendas.[85] Several climate change adaptation policies have been adopted for long-term management of floods, such as those regarding flood risk mitigation of the Thames Estuary over the 21st century.[86] Another example is the city of Malmö, Sweden, which has developed surface-level stormwater systems, green roofs, green walls, and green spaces to reduce flooding and recharge groundwater.[3]

It is important to emphasize that there is no "one size fits all" solution. The environmental issues that cities face vary widely, so cities have to find their own solutions and policies that respond best to their individual problems. For example, not all coastlines will experience the same sea level rise, as a consequence of variations in ocean volume, land-water-ice mass changes, tectonics, natural sediment compaction, occupation of floodplains, increased imperviousness, and aquifer dewatering.[10,88] For example, in the Holland delta, the flood protection system includes a series of dikes and storm surge barriers built to protect cities from one in 10,000-year flood events. Although this system requires regular high-level maintenance, it has proven to be very successful.[89]

Accurate predictions of the exposure of cities to flooding due to changes in land cover and climate are required in order to develop appropriate water management systems and flood resilience. However, despite efforts to control floods, the flood risk is increasing. One reason is that although flood forecasting models are constantly being improved, they remain partly inaccurate due to uncertainties in data processing.[90] Approaches combining remote sensing and flood modeling to quantify the influence of urban development and predicted climate change on the exposure of urban areas to flooding are suitable for large-scale mapping of the vulnerability of cities to climate change.[91] This approach facilitates generation of maps that take into account past, present, and future urban development and climate change, using a reliable method that is transferable across cities and countries.

Rapid urban development coupled with climate change have increased the flood risk and pose corresponding challenges to urban flood risk management, circumstances which require integration of new concepts such as urban resilience into flood management.[90] The urban resilience model considers the city as an entity composed of different elements such as population, companies, public infrastructure, housing, and networks in which the power of the city is based on its ability to provide a safe environment and efficiently support residents' activities.[92] This approach consists of preventive and emergency measures, including building flood-resilient buildings and adopting a system approach to flood risk management.[93]

Structural measures are often effective in controlling fluvial or riverine floods from upstream catchment areas. However, with the challenges arising from rapid urbanization and increasing frequency of precipitation extremes associated with climate change, it is becoming increasingly difficult to ensure protection of downstream urban areas.[67] Conventional drainage solutions are not designed to cope with such extreme events (climate change), and consequently other solutions for water conveyance and storage have to be found, such as the use of careful spatial planning.[75] One promising area is the growing suite of nature-based solutions (e.g., green roofs, raingardens) in response to a range of issues from flood risks and heatwaves to energy consumption and promotion of health and well-being.

5. Final considerations

Water management in urban areas remains one of the main challenges for society, given the trend for increasing urbanization and climate change projections. This is particularly relevant in developing countries with limited information on population data and lacking spatial planning. Expansion of impervious surfaces during urbanization often causes decreasing water infiltration and storage capacity, and increasing runoff and flood hazard. However, the magnitude of the impacts varies depending on flow connectivity, determined by the extent and pattern of impervious and pervious surfaces within the urban landscapes, and the artificial drainage systems installed. Modeling has been widely used to identify the hydrological processes in urban areas and support flood risk assessment, but there are still uncertainties in estimating water depth, flow velocity and the extent of flood-susceptible areas, issues which require further investigation. Flood risk assessment is the best basis to improve urban flooding conditions, through design and implementation of appropriate strategies, often reliant on structural measures (e.g., flood levees, embankments) which are generally considered effective in mitigating flood risk. Different flood risk measures should be established for different heavy precipitation scenarios to enable more differentiated protection, based on planning, flood prevention, and mitigation. A paradigm shift from flood protection to flood resilience approaches has been occurring in recent years. There is increasing interest in sustainable flood management, an approach that is reliant on providing opportunities for water interception, evapotranspiration, infiltration, and filtration, and thus reduced runoff and floods, mainly through vegetated surfaces. Different types of sustainable water management systems have been implemented at different scales in urban areas (e.g., green roofs, artificial wetlands), but their specific design, size, and location affect their capacity to cope with runoff from different storm events. Thus, although sustainable water measures are seen as a relevant complement to conventional drainage systems for effective flood mitigation, development of a stronger evidence base is key for successful mainstreaming of such measures. Besides flood risk mitigation, these measures often offer multifunctional spaces able to deliver a wide array of ecosystem services particularly relevant in urban areas (e.g., increased biodiversity, enhanced human satisfaction). These co-benefits should be considered when deliberating the relevance of sustainable water systems for flood protection. These are important aspects to consider for urban planning, given the complex environmental, social, and economic challenges and the limited space available to meet a wide range of needs. In fact, the involvement of multiple authorities controlling different aspects of urban areas is a major challenge to implementation of sustainable water systems for flood mitigation. Governance systems must be streamlined and coordinated to enhance sustainable flood risk mitigation and urban resilience.

Acknowledgments

This research was supported by the Bolin Centre for Climate Research, through the project "Nature-based solutions in Stockholm: Supporting adaptation to environmental challenges and enhancing urban resilience."

References

1. World Bank. *Urban population (% of total population)*. Washington, DC USA: The World Bank–Data; 2021. From the W.W.W. (https://data.worldbank.org/indicator/SP.URB.TOTL.IN.ZS) (Retrived 21 December 2021).
2. UNHABITAT. *Cities and climate change: global report on human settlements 2011*. London, UK and Washington, DC, USA: Earthscan Publications Ltd; 2011. p. 279.
3. European Union, UN-HABITAT. *The State of European Cities*; 2016. p. 2016. https://ec.europa.eu/regional_policy/sources/policy/themes/cities-report/state_eu_cities2016_en.pdf.
4. Zambon I, Rontos K, Serra P, Colantoni A, Salvati L. Population dynamics in southern Europe: a local-scale analysis, 1961–2011. *Sustainability* 2019;**11**(1):109.
5. Antrop M. Landscape change and the urbanization process in Europe. *Landsc Urban Plan* 2014;**67**(1–4):9–26.

6. UN HABITAT. *Urban planning responses in post crisis context.* Nairobi, Kenya: UN-HABITAT; 2019. p. 140.
7. Marsalek JB, Jiménez-Cisneros E, Malmquist PA, Karamouz M, Goldenfum J, Chocat B. Urban water cycle processes and interactions. In: *IHP-VI technical documents in hydrology no. 78, UNESCO, Paris, 2006*; 2006.
8. Galderisi A, Limongi G, Salata KD. Strengths and weaknesses of the 100 resilient cities initiative in Southern Europe: Rome and Athens' experience. *City Territory Architect* 2020;**7**:16.
9. IPCC. Climate change 2014: impacts, adaptation, and vulnerability. Part A: global and sectoral aspects. In: Field CB, Barros VR, Dokken DJ, Mach KJ, Mastrandrea MD, Bilir TE, et al., editors. *Contribution of working group II to the fifth assessment report of the intergovernmental panel on climate change.* Cambridge, UK: Cambridge University Press; 2014. p. 1132.
10. Yin J, Jonkman S, Lin N, Yu D, Aerts J, Wilby R, et al. Flood risks in sinking delta cities: time for a reevaluation? *Earth's Future Adv Earth's Space Sci* 2020. https://doi.org/10.1029/2020EF001614.
11. Hallegatte S, Green C, Nicholls RJ, Corfee-Morlot J. Future flood losses in major coastal cities. *Nat Clim Chang* 2013;**3**(9):802–6.
12. Jonkman SN. Global perspectives on loss of human life caused my floods. *Natural Hazard* 2005;**34**:151–75.
13. Mrozik K, Przybyła C, Pyszny K. Problems of the integrated urban water management. The case of the Poznań Metropolitan Area (Poland). *Ann Set Environ Prot* 2015;**17**:230–45.
14. Hrkić IZ, Kapović SM, Šumatić N, Ristić R, Marjanović-Balaban Ž. The role of plants in water regulation and pollution control. In: *The handbook of environmental chemistry.* Berlin, Heidelberg: Springer; 2021.
15. Rijke J, Brown R, Zevenbergen C, Ashley R, Farrelly M, Herk V, et al. Fit-for-purpose governance: a framework to operationalise adaptive governance. *Environ Sci Pol* 2012;**22**:73–84.
16. Walsh CL, Roberts RJ, Dawson JW, Hall A, Nickson HR. Experiences of integrated assessment of climate impacts, adaptation and mitigation modelling in London and Durban. *Environ Urban* 2013;**25**(2):361–80.
17. IPCC. Annex I: Glossary. In: Matthews JBR, Masson-Delmotte V, Zhai P, Pörtner H-O, Roberts D, Skea J, Waterfield T, editors. *Global warming of 1.5°C. An IPCC special report on the impacts of global warming of 1.5°C above pre-industrial levels and related global greenhouse gas emission pathways, in the context of strengthening the global response to the threat of climate change, sustainable development, and efforts to eradicate poverty.* IPCC; 2018.
18. Gasper R, Blohm A, Ruth M. Social and economic impacts of climate change on the urban environment. *Curr Opin Environ Sustain* 2011;**3**(3):150–7.
19. Brown A, Dayal A, Del Rio CR. From practice to theory: emerging lessons from Asia for building urban climate change resilience. *Environ Urban* 2012;**24**(2):531–56.
20. Ferrans P, Torres MN, Temprano J, Sánchez JR. Sustainable Urban Drainage System (SUDS) modeling supporting decision-making: a systematic quantitative review. *Sci Total Environ* 2022;**806**:150447.
21. Maidment DR. *Handbook of hydrology.* New York, NY: McGraw-Hill, Inc; 1993.
22. Duarte AC, Mateos L. How changes in cropping intensity affect water usage in an irrigated Mediterranean catchment. *Agric Water Manag* 2021;**260**:107274.
23. Kalantari Z, Ferreira CSS, Walsh RPD, Ferreira AJD, Destouni G. Urbanization development under climate change: hydrological responses in a peri-urban Mediterranean catchment. *Land Degrad Dev* 2017;**28**(7):2207–21.
24. Arnold CL, Gibbons CJ. Impervious surface coverage: the emergence of a key environmental indicator. *J Am Plan Assoc* 1996;**62**(2):243–58.

25. Yang D, Yang Y, Xia J. Hydrological cycle and water resources in a changing world: a review. *Geogr Sustain* 2021;**2**:115–22.
26. Caro-Borrero A, Carmona-Jiménez J, Rivera-Ramírez K, Bieber K. The effects of urbanization on aquatic ecosystems in peri-urban protected areas of Mexico City: the contradictory discourse of conservation amid expansion of informal settlements. *Land Use Policy* 2021;**102**:105226.
27. Ramier D, Berthier E, Andrieu H. The hydrological behaviour of urban streets: long-term observations and modelling of runoff losses and rainfall, runoff transformation. *Hydrol Process* 2011;**25**:2161–78.
28. Albrecht JC. *Alterations in the hydrologic cycle induced by urbanization in northern new castle county, Delaware: magnitudes and projections.* US Environmental Protection Agency Library Report; 1974. Number: DI-14-31-0001-3508; OWRR-A-017-DEL; W74-07729; OWRR-A-017-DEL(2).
29. Wijesekara GN, Gupta A, Valeo C, Hasbani J-G, Qiao Y, Delaney P, et al. Assessing the impact of future land-use changes on hydrological processes in the Elbow River watershed in southern Alberta. *Canada J Hydrol* 2012;**412–413**:220–32.
30. Barron OV, Barr A, Donn M. Effect of urbanisation on the water balance of a catchment with shallow groundwater. *J Hydrol* 2012;**485**:162–76.
31. Chu ML, Knouft JH, Ghulam A, Guzman JA, Pan Z. Impacts of urbanization on river flow frequency: a controlled experimental modelling-based evaluation approach. *J Hydrol* 2013;**495**:1–12.
32. Ferreira CSS, Walsh RPD, Steenhuis TS, Shakesby RA, Nunes JPN, Coelho COA, et al. Spatiotemporal variability of hydrologic soil properties and the implications for overland flow and land management in a peri-urban Mediterranean catchment. *J Hydrol* 2015;**525**:249–63.
33. Ferreira CSS, Moruzzi R, Isidoro JMGP, Tudor M, Vargas M, Ferreira AJD, et al. Impacts of distinct spatial arrangements of impervious surfaces on runoff and sediment fluxes from laboratory experiments. *Anthropocene* 2019;**28**:100219.
34. Yang JL, Zhang GL. Water infiltration in urban soils and its effects on the quantity and quality of runoff. *J Soils Sediments* 2011;**11**(5):751–61.
35. Duarte AC, Ferreira C, Vitali G. Use of simulation models to aid soil and water conservation actions for sustainable agro-forested systems. In: Jhariya MK, Meena RS, Banerjee A, Meena SN, editors. *Natural resources conservation and advances for sustainability.* Amsterdam, Netherlands: Elsevier; 2021. p. 389–412.
36. Boyd MJ, Bufill MC, Knee RM. Pervious and impervious runoff in urban catchments. *Hydrol Sci J* 1993;**38**:463–78.
37. Jacobson CR. Identification and quantification of the hydrological impacts of imperviousness in urban catchments: a review. *J Environ Manag* 2011;**92**:1438–48.
38. Ferreira CSS, Walsh RPD, Steenhuis TS, Ferreira AJD. Effect of peri-urban development and lithology on streamflow in a Mediterranean catchment. *Land Degrad Dev* 2018;**29**:1141–53.
39. Soulsby C, Tetzlaff D, Rodgers P, Dunn S, Waldron S. Runoff processes, streamwater residence times and controlling landscape characteristics in a mesoscale catchment: an initial evaluation. *J Hydrol* 2006;**325**:197–221.
40. Berezowski T, Chormanski J, Batelaan O, Canters F, de Voorde TV. Impact of remotely sensed land-cover proportions on urban runoff prediction. *Int J Appl Earth Obs Geoinf* 2012;**16**:54–65.
41. Wang Z, Wua QJ, Wua L, Ritsema CJ, Dekker LW, Feyen J. Effects of soil water repellency on infiltration rate and flow instability. *J Hydrol* 2000;**231–232**:265–76.
42. Zhang Y, Shuster W. Impacts of spatial distribution of impervious areas on runoff response of hillslope catchments: simulation study. *J Hydrol Eng* 2014;**19**(6):1089–100.

43. Pappas EA, Smith DR, Huang C, Shuster WD, Bonta JV. Impervious surface impacts to runoff and sediment discharge under laboratory rainfall simulation. *Catena* 2008;**72**:146–52.
44. Reed SM, Fulton R, Zhang Z, Guan S. Use of precipitation forecasts to drive a distributed hydrologic model for flash flood prediction. In: *American meteorological society 20th conference on hydrology, Atlanta, GA*; 2006.
45. Foster S, Morris B, Lawrence A, Chilton J. Groundwater impacts and issues in developing cities. In: Chilton J, editor. *Groundwater in the urban environment. Proceedings of the XXVII IAHS Congress on Groundwater in the Urban Environment, Nottingham, UK. IAH*, vol. 21; 1999. p. 3–16.
46. Wheater H, Evans E. Land use, water management and future flood risk. *Land Use Policy* 2009;**26**:251–64.
47. Navratil O, Breil P, Schmitt L, Grosprêtre L, Albert MB. Hydrogeomorphic adjustments of stream channels disturbed by urban runoff (Yzeron River basin, France). *J Hydrol* 2013;**485**:24–36.
48. Liguori S, Rico-Ramirez MA, Schellart A, Saul AJ. Using probabilistic radar rainfall nowcasts and NWP forecasts for flow prediction in urban catchments. *Atmos Res* 2012;**103**:80–95.
49. Vasiloff S, Seo DJ, Howard K, Zhang J, Kitzmiller D, Mullusky M, et al. Improving QPE and very short term QPE. *Bull Am Meteorol Soc* 2007;1899–911.
50. Elliott AH, Trowsdale SA. A review of models for low impact urban stormwater drainage. *Environ Model Softw* 2007;**22**:394–405.
51. L.A. Rossman, Storm Water Management Model User's Manual, Version 5.0, 2010, U.S. Environmental Protection Agency. Office of Research and Development. EPA/600/R-05/040.
52. Purdue-University. *L-THIA-LID*; 2016. https://engineering.purdue.edu/~lthia/.
53. Fletcher TD, Andrieu H, Hamel P. Understanding, management and modelling of urban hydrology and its consequences for receiving waters: a state of the art. *Adv Water Resour* 2013;**51**:261–79.
54. Deletic A, Dotto CS, McCarthy DT, Kleidorfer M, Freni G, Mannina G. Assessing uncertainties in urban drainage models. *Phys Chem Earth* 2011;**42–44**:3–10.
55. Hawken S, Sepasgozar S, Prodanovic V, Jing J, Bakelmun A, Avazpour B, et al. What makes a successful Sponge City project? Expert perceptions of critical factors in integrated urban water management in the Asia-Pacific. *Sustain Cities Soc* 2021;**75**:103317.
56. Poku-Boansi M, Amoako C, Owusu-Ansah J, Cobbinah P. What the state does but fails: exploring smart options for urban flood risk management in informal Accra, Ghana. *City Environ Interact* 2020;**5**, 100038.
57. Chan F, Yang L, Scheffran J, Mitchell G, Adekola O, Griffiths J, et al. Urban flood risks and emerging challenges in a Chinese delta: the case of the Pearl river delta. *Environ Sci Policy* 2021;**122**:101–15.
58. Silva L, Alencar M, Almeida A. Multidimensional flood risk management under climate changes: bibliometric analysis, trends and strategic guidelines for decision-making in urban dynamics. *Int J Disaster Risk Reduct* 2020;**50**:101865.
59. Ciampa F, Seifollahi-Aghmiuni S, Kalantari Z, Ferreira CSS. Flood mitigation in mediterranean coastal regions: problems, solutions, and stakeholder involvement. *Sustainability* 2021;**131**:474.
60. Rahmati O, Darabi H, Panahi M, Kalantari Z, Naghibi SA, Ferreira CSS, et al. Development of novel hybridized models for urban flood susceptibility mapping. *Sci Rep* 2020;**10**:12937.
61. Ferreira CSS, Walsh RPD, Kalantari Z, Ferreira AJD. Impact of land-use changes on spatiotemporal suspended sediment dynamics within a peri-urban catchment. *Water* 2020;**12**:665.

62. Ferreira CSS, Mourato S, Ksanin-Grubin M, Ferreira AJD, Destouni G, Kalantari Z. Effectiveness of nature-based solutions in mitigating flood hazard in a Mediterranean PeriUrban catchment. *Water* 2020;**12**:2893.
63. Chen J, Huang G, Chen W. Towards better flood risk management: assessing flood risk and investigating the potential mechanism based on machine learning models. *J Environ Manag* 2021;**293**:112810.
64. Rana I, Asim M, Aslam A, Jamshed A. Disaster management cycle and its application for flood risk reduction in urban areas of Pakistan. *Urban Clim* 2021;**38**:100893.
65. Gobel P, Stubbe H, Weinert M, Zimmermann J, Fach S, Dierkes C, et al. Near-natural stormwater management and its effects on the water budget and groundwater surface in urban areas taking account of the hydrogeological conditions. *J Hydrol* 2004;**299**:267–83.
66. Rubinato M, Nichols A, Peng Y, Zhang J, Lashford C, Cai Y, et al. Urban and river flooding: comparison of flood risk management approaches in the UK and China and an assessment of future knowledge needs. *Water Sci Eng* 2019;**12**(4):274–83.
67. Chan F, Griffiths J, Higgitt D, Xu S, Zhu F, Tang Y, et al. "Sponge City" in China—a breakthrough of planning and flood risk management in the urban context. *Land Use Policy* 2018;**76**:772–8.
68. Herk S, Zevenbergen C, Ashley R, Rijke J. Learning and Action Alliances for the integration of flood risk management into urban planning: a new framework from empirical evidence from The Netherlands. *Environ Sci Policy* 2011;**14**:543–54.
69. Sheehan J, Brown J. Flood risk management: property rights-focussed instruments in Australia. *Environ Sci Policy* 2021;**119**:12–7.
70. Li X, Erpicum S, Mignot E, Archambeau P, Pirtton M, Deals B. Influence of urban forms on long-duration urban flooding: laboratory experiments and computational analysis. *J Hydrol* 2021;**603**:127034.
71. Disse M, Johnson T, Leandro J, Hartmann T. Exploring the relation between flood risk management and flood resilience. *Water Security* 2020;**9**:100059.
72. Kuller M, Bach P, Ramirez-Lovering D, Deletic A. Framing water sensitive urban design as part of the urban form: a critical review of tools for best planning practice. *Environ Model Softw* 2017;**96**:265–82.
73. Darnthamrongkul W, Mozingo L. Toward sustainable stormwater management: understanding public appreciation and recognition of urban Low Impact Development (LID) in the San Francisco Bay Area. *J Environ Manag* 2021;**300**:113716.
74. Tang S, Jiang J, Zheng Y, Hong Y, Chung E, Shamseldin AY, et al. Robustness analysis of storm water quality modelling with lid infrastructures from natural event-based field monitoring. *Sci Total Environ* 2021;**753**:142007.
75. Nieuwenhuis E, Cuppen E, Langeveld J, Bruijn H. Towards the integrated management of urban water systems: conceptualizing integration and its uncertainties. *J Clean Prod* 2021;**280**:124977.
76. Mai T, Mshtaq S, Reardon-Smith K, Webb P, Stone R, Kath J, et al. Defining flood risk management strategies: a systems approach. *Int J Disaster Risk Reduct* 2020;**47**:101550.
77. Löwe R, Mair M, Pedersen A, Kleidorfer M, Rauch W, Arnbjerg-Nielsen K. Impacts of urban development on urban water management—limits of predictability. *Comput Environ Urban Syst* 2020;**84**:101546.
78. Pinto L, Ferreira CSS, Pereira P. Environmental and socioeconomic factors influencing the use of urban green spaces in Coimbra (Portugal). *Sci Total Environ* 2021;**792**:148293.
79. Raska P, Bezak N, Ferreira CSS, Kalantari Z, Banasik K, Bertola M, et al. Identifying barriers for nature-based solutions in flood risk management: an interdisciplinary overview using expert community approach. *J Environ Manag* 2022;**310**:114725.
80. Solomun MK, Ferreira CSS, Barger N, Tosic R, Eremija S. Policy framework on land degradation from a post conflict perspective of Bosnia and Herzegovina. *Land Degrad Dev* 2020;1–10.

81. Johnston P. *Why private residential water recycling has stalled in Sydney*; 2019 https://thefifthestate.com.au/urbanism/environment/why-private-residential-water-recycling-has-stalled-in-sydney/. [Accessed November 2021].
82. Dorst H, Jagt A, Runhaar H, Raven R. Structural conditions for the wider uptake of urban nature-based solutions—a conceptual framework. *Cities* 2022;**116**:103283.
83. Kabisch N, Frantzeskaki N, Pauleit S, Naumann S, Davis M, Artmann M, et al. Nature-based solutions to climate change mitigation and adaptation in urban areas: perspectives on indicators, knowledge gaps, barriers, and opportunities for action. *Ecol Soc* 2016;**21**(2):39.
84. UN Water/WWAP (United Nations World Water Assessment Programme). *The United Nations world water development report 2018: nature-based solutions for water.* Paris: UNESCO; 2018.
85. GAR—Global Assessment Report on Disaster Risk Reduction. *United Nations office for disaster risk reduction*; 2019. https://gar.undrr.org/.
86. Ristić RK, Solomun MM, Ždrale I, Radić S, Polovina B, Milćanović S. Healthy soils—healthy people, reality of Balkan region. In: *The soil-human health nexus.* New York, USA: Taylor and Francis group; 2021.
87. EEA. *Land and soil in Europe: why we need to use these vital and finite resources sustainability.* European Environment Agency; 2019. https://doi.org/10.2800/779710.
88. Abadie LM, Jackson LP, de Murieta ES, Jevrekeva S, Galarrage I. Comparing urban coastal flood risk in 136 cities under two alternative sea-level projections: RCP 8.5 and expert opinion-based high-end scenario. *Ocean Coast Manag* 2020;**193**:105249. https://doi.org/10.1016/j.ocecoaman.2020.105249.
89. Yin J, Yu D, Yin Z, Wang J, Xu S. Modelling the anthropogenic impacts on fluvial flood risks in a coastal mega-city: a scenario-based case study in Shanghai, China. *Landsc Urban Plan* 2015;**136**:144–55.
90. Lhomme S, Serre D, Diab Y, Laganier R. Analyzing resilience of urban networks: a preliminary step towards more flood resilient cities. *Nat Haz Earth Syst Sci* 2013;**13**:221–30.
91. Skougaard Maspersen P, Høegh Ravn N, Arnjerg-Nielsen K, Madsen H, Drews M. Influence of urban land cover changes and climate change for the exposure of European cities to flooding during high-intensity precipitation. *Proc IAHS* 2015;**370**:21–7.
92. Lhomme S, Serre D, Diab Y, Laganier R. *GIS development for urban resilience, Sustainable City 2010, 14–16 April 2010.* Spain: La Coruna; 2010.
93. Pasche E, Geisler TR. New strategies of damage reduction in urban areas proned to flood. In: Szollosi-Nagy A, Zevenbergen C, editors. *Urban flood management.* WIT Press; 2005. p. 101–17.

CHAPTER FOUR

Land and flood protection: A strategic planning experience in Franciacorta (Lombardy, Italy)

Michele Pezzagno*
Department of Civil, Environmental, Architectural Engineering and Mathematics, DICATAM, University of Brescia, Brescia, Italy
*Corresponding author: e-mail address: michele.pezzagno@unibs.it

Contents

Abstract

The contribution proposes a governance approach on strategic spatial planning devoted to land protection. The case study involves 22 municipalities located in Franciacorta (about 200,000 inhabitants and 300 sq. km), which is also an important wine-area in northern Italy. It is a story of partial failure that makes evidence of the difficulties of engagement processes on land and flood protection that need inter-institutional approaches and collaborative attitudes among municipalities to reach a more sustainable land-use practices and high quality of life for their citizens. The study underlines that strong institutions play a fundamental role to lead communities and their transformation capacity to use appropriate and inclusive governance mechanisms in transition territories.

Keywords: Strategic plan, Governance process, Spatial planning, Flooding, Land protection

Advances in Chemical Pollution, Environmental Management and Protection, Volume 8
ISSN 2468-9289
https://doi.org/10.1016/bs.apmp.2022.10.014

1. Introduction

The European Floods Risk Directive (2007/60/EC) is the first European Directive that deals specifically with floods. In the first step it requires EU Member States to undertake a preliminary assessment of flood risks, in the second it requires to prepare flood hazard and risk maps and in the third one, it requires drawing up Flood risk management plans which were expected to be ready in 2015. These plans "may also include the promotion of sustainable land-use practices (e.g., nature-based solutions, de-sealing, flood retention basins, etc.), improvement of water retention as well as the controlled flooding of certain areas in the case of flood event" (art. 7.3. 2007/60/EC).[1]

In Italy, the Directive was transposed in the National legislation by Legislative Decree 152/2006 and the previous Law 183/1989 on the soil defense was formally abrogated. As argued by Vitale and Meijerink,[2,3] Italian flood risk policies are still characterized by a sectoral engineering approach rather than a risk-based approach. This technical domain is primarily dominated by experts in building flood control infrastructure, while spatial policies for flood risk reduction are poorly developed. Long term strategic investments, spatial policies, and visions are missing and the occurrence of shock events, such as the one in Sarno (Law267/98) (5th May 1998), in which exceptional rainfalls caused a catastrophic landslide that involved five municipalities, and in Soverato (Law.365/00) (8th–10th September 2000), where exceptional rainfalls caused the overflow of the Beltrame Stream and a dam break, triggered ad hoc policy responses with a general lack of attention on core policies. Flood Risk Directive application requires profound changes in our management paradigm.[4] Institutional path dependencies have generated hard interdependence of system elements, such as institutional design, technical infrastructure, knowledge, and distribution of power, that guarantee the functioning of a system and the convergence of expectations of actors.[5] Furthermore, targeted state financial resources in ad hoc sectoral policy such as "Italia Sicura" (2014) and "Proteggi Italia" (2019) boost a technocratic approach instead of a holistic approach. Italian decentralization of the state power started after the II World War and has been characterized by a partial autonomy of the region where the state maintains or regains the power. Also, concerning flood risk policies, the effect is a power-play between the national and regional levels that generates a lack of coordination with progressive overloading of local authorities.[6]

In a context in which collaborative bottom-up initiatives are unusual and in which an active role of local officials tends to underestimate and even deny local risks,[2,3] this contribution describes a pilot experience of integration of spatial planning with flood risk management conducted during the strategic planning of the Franciacorta geographical area (Strategic Plan/*Piano territoriale regionale d'Area*, according to Region Lombardy Law n.12/2005).

Unlike previous experiences, the process described in this chapter is based on an inter-institutional approach supported by the university to try to overcome the barriers related to institutional path dependencies and turn them into collaborative strategic spatial planning through a risk-based approach.

2. Study area

The Franciacorta wine area lies at the heart of Lombardy Region (Italy). It is a suburban area close to the city of Brescia, that is the chief city of the *Province of Brescia* and it is one of the major cities of the region (NUT2). The province is located in the north-eastern side of the region and represents the most extended province of Lombardy (4785.62 km^2) and the second province per number of inhabitants (1,254,322 inhabitants, 262 inhab/km^2)[7] (Fig. 1). It stretches from the Alps to the Po plain, featuring mountainous landscapes in the north, highly urbanized area in the middle,

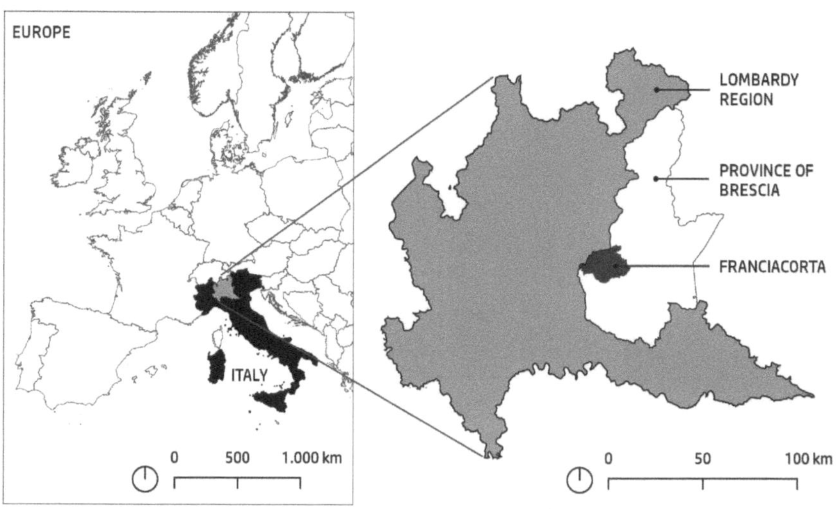

Fig. 1 Location of Italy, the Lombardy region, the province of Brescia and the Franciacorta study area.

and rural areas in the south. Almost 30% of this area is at risk of flooding. The Franciacorta wine area extends for about 200 km^2 and is highly located between the sub-mountainous and the high Po River alluvial plain bands, in which urban settlements generated a dense urbanized net that has fragmented the agricultural landscape of the plain. Between the climate, topography, and historical significance, Franciacorta is a wine-producing area with a relatively young history—the first official Franciacorta sparkling wine was produced in 1961.

Franciacorta wine is a sparkling wine of exceptional quality similar in style to French champagne. This wine production made exclusively in this region has been disciplined since 1967 (rif. DPR 21.07.1967 Gazzetta Ufficiale n° 290 del 21/08/1967). The region, infused with cultural significance and a thriving enological tradition, has quickly established as one of the country's *premiere* wine regions thanks to its meticulous production standards and constant investment in high-quality production.[8]

Grape varieties are not only strictly related to Franciacorta high-quality wine production (Protected Designation of Origin—PDO) (Regulation (EU) n. 1308/2013) representing more than 3200 ha of the land cover.[9] The traditional cultivation based on terraces is one of the most important characteristics of a stunning landscape of vineyard-clad hills, charming Italian countryside, hundreds-year-old architecture, and surrounded by natural wonders of rivers, lakes and mountains.

Despite these strong agricultural vocations, the territory, from the post-war period to today, is undergoing an intense urbanization process, also linked to the socio-economic changes that have occurred at the national level. Today, as a result, the area appears partially urbanized and, since it is close to the main nodes, looks like a strongly anthropized territory subject to phenomena of peripheralization from the city of Brescia. Along the main infrastructures, the presence of productive and commercial activities with a preponderant industrial component is remarkable.[10] The overall demographic size of the area is almost 200,000 inhabitants, representing 15% of the provincial population of Brescia.[11]

3. Toward the definition of a Strategic Plan for Franciacorta area

The collaborative path among the municipalities of Franciacorta region began in 2007 and can be framed as an experience of spontaneous governance. The municipalities recognizing the potentiality of a strong and aggregating identity related to wine production worked together at the Sustainable Franciacorta project (2007) and at the Land of Franciacorta project (2011).

Despite the overall awareness of the opportunity deriving from a shared idea of socio-economic development and a common strategy, over the years municipalities have implemented very different planning choices. The result is the progressive alteration of rural areas, soil degradation, land fragmentation, loss of values (e.g., cultural and heritage values, landscapes, etc.) and ecosystem services (pest control, disease control, etc.)[12] also related to the increasing soil sealing caused by urban sprawl. Furthermore, due to increasing land-take and soil sealing damages, flooding events are growing, making evidence of increasing risk conditions in a region where there are already several vulnerable areas both from a hydrogeological and hydraulic point of view.

Landscape change and territorial identity loss represents a potential damage to Franciacorta sparkling wine *brand* all over the world. Winescape is intended as ambient conditions, spatial layout, functionality of the territory in which wine is produced, and it influences the mood and the attitude of customers.[13,14] It is also well known that tourists do not plan visits to a wine region based only on the presence of wineries and wine products they seek a wide variety of cultural tourism and outdoor recreation supported by the high quality of the territory in which their experience takes place. For this reason, wine producers need a strong link with tourism to connect local products, countryside, traditions, and cultural values but also to place emphasis on the uniqueness of the territory.[15,16]

Antrop[17] underlines that the rural landscape is the result of specific choices in terms of both local governance and economic development policies. In the context of Franciacorta, the Franciacorta wine consortium—which represents over 100 wineries that produce Franciacorta sparkling wines—represents one of the main sponsors of the governance process of the area and supports collaborative high quality territorial policies able to promote Franciacorta sparkling wine, alongside which there are also other high-quality products.

The mayors of Franciacorta in 2013 signed a memorandum of understanding (MOU) requiring Region Lombardy Strategic planning action for this geographical area. The primary object of MOU is to retrain the landscape and improve the quality of life for Franciacorta inhabitants. Mayors engaged Region Lombardy on Franciacorta future as this region faced with a territory characterized by products that are officially certified not only as

Controlled and Guaranteed Designation of Origin (DOCG), that identifies the highest level of protection in the Italian wine classification (the top wines of the PDO European classification), but also by products of Protected Geographical Indication (PGI), such as oil (*Olio extravergine dei laghi lombardi, Olio Extravergine del Sebino*) and cheese (gorgonzola, grana padano, taleggio, provolone valpadana) (Fig. 2).

The Strategic Plan of the Franciacorta geographical area (Area Regional Territorial Plan/*Piano territoriale regionale d'Area*, according to Region Lombardy law 12/2005) involves 22 municipalities. Eighteen municipalities are the ones of the Franciacorta wine area disciplined since 1967 (rif. DPR 21.07.1967 Gazzetta Ufficiale n° 290 del 21/08/1967) denominated "Core" and four municipalities are the ones bordering with the Franciacorta wine area denominated "Buffer" (Fig. 3).

After the Plan approval, Region Lombardy has also launched a process to support the governance aimed at increasing the capacity of local authorities to converge toward common reflections (and choices) on spatial planning opportunities, especially concerning soil protection and hydrogeological risk reduction in those areas in which the urbanization process undergone during recent years has been particularly strong.

Fig. 2 High quality agricultural production and urban development in Franciacorta.

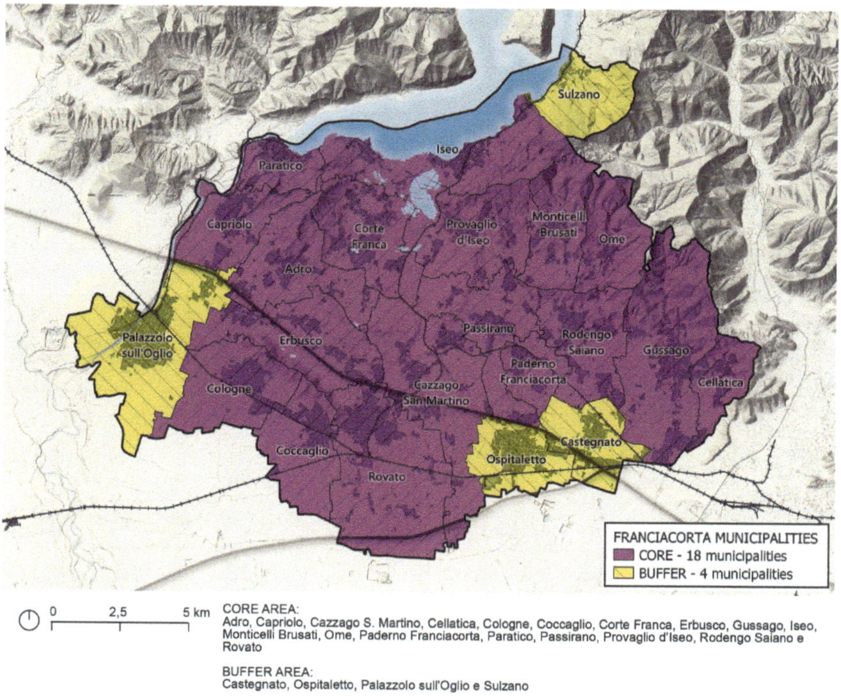

Fig. 3 The 22 municipalities involved in the Strategic Plan of the Franciacorta geographical area.

4. Land-use analysis and hydrogeological risk assessment in Franciacorta

Franciacorta is characterized mostly by a series of dense urban agglomerations surrounded by discontinuous urbanized fabric and farm-houses that determined urban sprawl phenomena. This area covers 6.6% of the Province of Brescia. Although the population of Franciacorta region has been slightly decreasing in recent years (−0.49% between 2016 and 2022), the overall demographic asset (e.g., the population density and the population variation between 2001 and 2016) confirms the attractiveness of the area (Table 1).

Focusing on the "Core" of Franciacorta, which is of strategic interest for the wine producers, the urbanized land is about 63.799 sq. km (data from Mosaico dei piani della provincia di Brescia, 2015) and the main land uses related to the anthropized area are residential 43%; infrastructural

Table 1 Demographic assets of Franciacorta compared to the Province of Brescia.[7,11]

Demographic asset	**Franciacorta**	**Province of Brescia**
Number of municipalities	22	205
Area (km^2)	316	4786
Average area of municipalities (km^2)	14.4	23.1
Population (at 1.1.2022)	193,848	1,254,322
Population density (inhab./km^2)	613	262.08
Population variation (2001–2016)	+20%	+13%
Population variation (2016–2022)	−0.49%	−0.85%

Table 2 Land-take phenomena in Franciacorta.

Urbanized land	**sq. km**	**%**
Historical/consolidated urban settlement	18.170	83.32%
Transformation area—reconstructing	0.059	0.27%
Transformation area—conversion	0.321	1.47%
Transformation area—land-take	3.257	14.94%

Data from "Mosaico dei piani. Province of Brescia." Data updated in 2015.

(including plant, services, urban facilities) 27%, productive/commercial 21% (QC0—*Relazione del Piano territoriale regionale d'area della Franciacorta, Regione Lombardia*) (Table 2).

The cultivated land in the "Core" is about 14,066 ha subdivided into 28,255 rural parcels. The medium parcel is less than a hectare (0.49 ha), determining a high fragmentation of cultivations. The cropping system is based on grain maize (26.4%), vine (22.9%), grass plants (12.3%) and waxy maize (11.1%) which represent more than 70% of the cultivations.

In 2015, Region Lombardy was drawing up the Flood Risk Management Plan according to the 2007/60/EC directive and a collaborative process opportunity was foreseen to introduce an updated hydrogeological risk assessment in the Strategic plan of Franciacorta.

The flooding areas identified by the regional Flood Risk Management Plan approved in 2015 (*Piano di Gestione del Rischio Alluvioni—PGRA*, 2015) have been introduced using existing studies including, in particular, the Hydrogeological Asset Plan, the Feasibility Studies of the Po River Basin Authority, the contents of municipal geological studies and the data provided by Regional Union of Irrigation Reclamation Land Improvements

(Unione Regionale Bonifiche Irrigazioni Miglioramenti Fondiari URBIM Lombardia) relating to the areas historically flooded. In particular, the municipal geological studies are quite heterogeneous in the timeline and elaborated according to different risk definitions related to the succeeding regional laws.

In Franciacorta two Areas of Potential Significant Flood Risk (Aree a Rischio Significativo—ARS) at Regional Level (RL) were drawn (Fig. 4) (PGRA, 2015). The first involves the municipalities of Cellatica, Gussago, Rodengo Saiano which are located in the east part of the Franciacorta Core. In this case, the risk is caused by Gandovere, Canale and Livorna torrents (ASR19 of PGRA, 2015). The second involves the municipality of Palazzolo sull'Oglio which is located in the west part of the Franciacorta Buffer area (ASR21 of PGRA, 2015). These areas are characterized by high flood flows, significant extension of flood affected areas, involving residential and productive settlements, strategic infrastructures and main communication routes. Concerning the assessment of the hydrological risk of the Franciacorta Core in relation to ASR19, already existing hydrogeographic studies of the sub-basin have also been used to better understand the critical issues related to the minor water network.

Land-use choices can also increase the risk of creating situations of instability (landslides, erosion, etc.), even of considerable importance, variable according to local geological characteristics (PGRA, 2015).

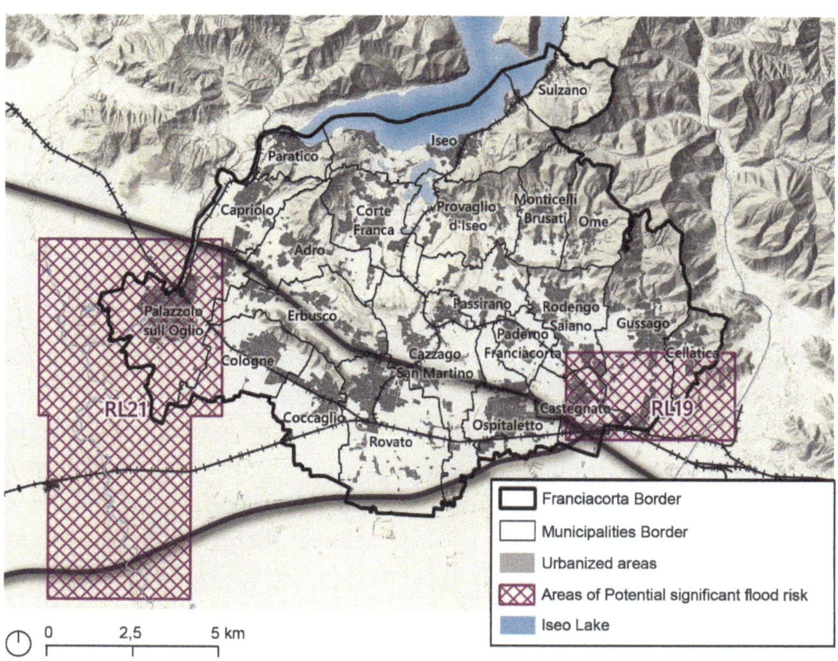

Fig. 4 Areas of potential significant risk at regional level in Franciacorta (PGRA, 2015).

The main critical issues for Franciacorta geographical area are related to the possibility of anthropic alteration of the morphology and the state of naturalness of the places through anthropogenic interventions determined by the exploitation of mountain resources (hydroelectric plants, power lines) with obvious traces of conflict with the natural context.

The landscape is compromised by all the transformations carried out on the slope of Franciacorta hills and to the marked visual exposure of the buildings arranged on steep slopes.

All these critical issues are related to the municipalities' inability in long-time maintenance of a common strategy and perspective on territorial development. It is difficult to understand why intensities of conflict and cooperation between municipalities and stakeholders have changed over time, however, starting from the position argued by Mirumachi and Allan,[18] it is evident that even at the local level dynamic bordering relations between municipalities are characterized by varying intensities of co-existing conflict and cooperation. Local power asymmetries in which some municipalities try to gain hegemony over the others negatively affect the decision-making slowing down the processes. In this situation, it is clear the importance of political leadership assumed by the Lombardy Region to push the outcomes of the strategic planning activities in the right direction.

5. The Strategic Plan of Franciacorta (Piano territoriale regionale d'Area della Franciacorta)

According to Ostrom research (2005, 2007), three types of exogenous factors that influence the action arena were taken into consideration during the Strategic Plan drawing: physical/material conditions, attributes of the community, and rules-in-use.[2,3] In Franciacorta the respective locations of municipalities create different options and dependencies in relation to the PDO/DOCG boundary (physical/material conditions). The strong local identity of the communities is characterized by different preferences, values, knowledge, and beliefs in achieving certain outcomes, influencing the consideration of the other municipalities involved in the planning action (attributes of the community). Traditionally every municipality develops its local masterplan and the rules foreseen to manage territorial development are heterogeneous.[19] The timeline and the agenda of the local plans are not aligned (rules-in-use) (Lombardy Region Law n. 12/2005).

Starting from this awareness the inter-institutional process was set up to accompany the drawing of the Strategic Plan for Franciacorta.

At the Regional level, an inter-directorate working group was officially instituted by Lombardy Region (Decree n. 6197, 22/07/2015, ID Working Group on "*Piano Territoriale Regionale d'Area Franciacorta*"), while at the local level a Co-planning Forum was established to engage all the 22 municipalities in the strategic choices.

After 2 years of participatory and collaborative design process, the output is a shared vision for Franciacorta geographical area focused on high-quality cultivation protection and hydrogeological risk assessment (Fig. 5), soft mobility development, and landscape restoration.

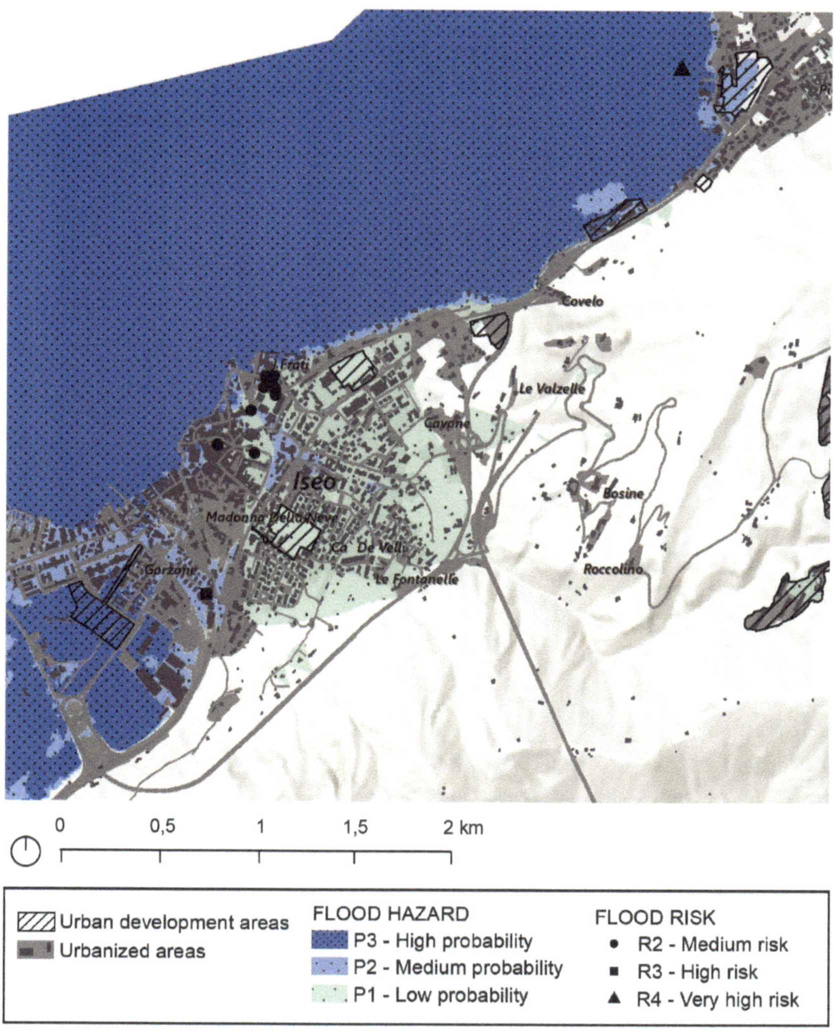

Fig. 5 Food risk and urban development in Franciacorta.

The Strategic Plan approved in 2017 (DCR n. X/1564—18 July 2017) therefore proposes new solutions for a territorial organization aimed at reducing the consumption of land, favoring soil protection through hydrogeological risk assessment, and enhancing cultural and landscape identities.[20]

The Plan aims to specify objectives and development strategies and outline a strategic reference scenario, as a guide to the coordinated and effective action of all the territorial programming of the subjects involved. It is the territorial governance tool that potentially allows the implementation of an effective synergy among the strategies of economic, social development and safeguarding the environmental sustainability of the territory involved, in order to harmonize policies, programs and projects, in the awareness that the environment is not only a matter of conservation but also of active intervention, despite many human activities aimed at development can have negative impacts. The Strategic Plan in its planning toolkit proposes a framework based on the dissemination of policy recommendations, spatial design criteria, guidelines for transformation areas and common rules to be directly applied at the local level by municipalities.[21,22]

6. Supporting governance and raising awareness

Over the course of the Co-planning events (more than 40 meetings between 29 January 2015 and 26 May 2016), organized during the Strategic Plan drawing, it was evident that several stakeholders were still promoting a short-sighted and speculative attitude favoring urban sprawl instead of land protection, without understanding that new planning principles based on de-sealing, restoration and reuse are needed to recover the incorrect past planning choices and improve the coexistence between urban activities and rural vocation of the area.

Therefore in 2018, after the Strategic Plan approval, Region Lombardy set up an accompanying process to increase the governance capacity of municipalities and consolidate the spatial planning vision for Franciacorta. The research group proposed to organize targeted events for Franciacorta local politicians and officials, in place of a more general involvement of stakeholders. From the research point of view, mayors seem to be unconscious of the complexity of the governance process and undervalue the internal barriers. Meanwhile, local officials seem to be worried about the capacities needed to manage the application of the spatial planning vision at the local level. The strategy for Franciacorta is holistic but needs a very technical approach to be managed and to reach the foreseen sustainable assessment of the territory.

The representatives of local administrations (mayors and municipal technicians—i.e., mainly architects and engineers employed in the technical offices of the local administration) were therefore involved in inter-institutional activities through questionnaires and co-design laboratories.

The questionnaire was devoted to better understanding the current state of the already existing different forms of governance in Franciacorta. Deepening if there are real collaborative or only opportunistic alliances among municipalities and if they are the result of a lack of funding (for ex: in services management) or based on common feelings in urban and territorial policies.

The laboratories were organized to pursue a shared representation to be used in the local urban planning tools of the municipalities of Franciacorta. In particular, Local officials' involvement through active research (a technician per each municipality, a technician from the Province of Brescia and a technician from Lombardy Region), was identified as key to facing the complex and potentially conflict-ridden process during the drawing of the general Master Plan (*Piano Regolatore Generale-PRG*) at the municipal level.

The goal of the whole participatory activities was to verify the real political attitude of the municipalities to promote a new common vision for the future of Franciacorta and the effective desire to reach technical-institutional coordination.

6.1 Questionnaire

Mayors' involvement was carried out through a questionnaire to assess the already existing collaboration among the 22 municipalities of Franciacorta. The survey, finalized by the research group, consisted of both open and closed questions and was spread out before starting with the laboratories attended by local officials. The research group explicitly engaged the 22 mayors in filling it, therefore a total of 22 questionnaires were delivered. The structure of the questionnaire is shown below (Fig. 6).

Part III of the questionnaire was devoted to understanding which actions of the Strategic Plan were considered a priority by the mayors. Only 12 mayors filled out the questionnaire: 6 mayors were not able to give a priority, while the other 6 were able to prioritize the actions considering soil protection and hydrogeological risk assessment as "the priority," while common rules for landscape restoration seemed to be the less important (Table 3).

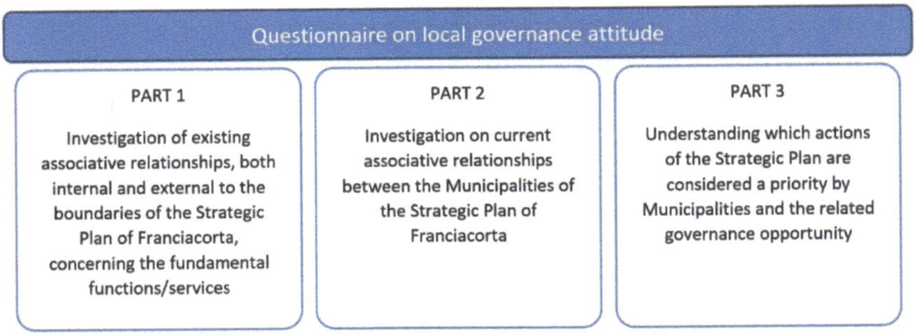

Fig. 6 Structure of the questionnaire on local governance attitude in Franciacorta (2018).

Table 3 Answer of part III of the questionnaire on local governance attitude in Franciacorta.

	Which is the priority action to be implemented by the Strategic Plan of Franciacorta?	working together on soil protection and hydrogeologica l risk assessment	working together on common rules for landscape restoration	acting together for soft mobility development
Municipality	Adro			
	Capriolo			
	Castegnato			
	Cellatica			
	Erbusco			
	Gussago			
	Iseo			
	Palazzolo sull'Oglio			
	Passirano			
	Provaglio d'Iseo			
	Rovato			
	Sulzano			
		75%	58%	67%

6.2 Laboratories

The laboratories aimed mainly at: involving the 22 municipalities in co-creating a landscape map of the Franciacorta Region, training the municipal technicians on the use of specific softwares, and studying the Land Use Map. Therefore, three thematic all-day-long laboratories were organized on the main actions foreseen in the Strategic Plan and all the local officials of the municipalities were invited to attend them (May, June, and October 2018).

According to the general aim of the process, the first laboratory was on "a shared legend for soil sealing" and worked on two different maps to represent both quantitative and qualitative aspects of soil sealing phenomena according to Regional Law of Lombardy 31/2005 criteria. Before starting the activity, the research group presented the data related to Franciacorta soil sealing and the recurrent land take phenomena. The first result of the laboratory is a co-production of maps based on the mapping criteria foreseen in the Regional Development Plan (*Piano Territoriale Regionale*). As a second result, the laboratory raised the awareness of the problem of vineyards consumption due to the land taken allowed by the local plans. More than 100,000 sq. m of vineyards are potentially lost in Franciacorta due to the building areas provided at the local level due to the general Master Plan rules.

The second laboratory was on "a shared legend on eco-landscape" putting in evidence high protected sites (e.g., Natura 2000 Network, Sites of Community Importance, National protected areas, Landscapes and cultural heritage protected by National Laws, etc.) as well as the sites less protected but of local interest to preserve the local landscape and local ecological network.

The work on the eco-landscape map makes evidence of the importance of common rules, needed by bordering municipalities, to protect the landscape (for example common rules for Montorfano hill in Franciacorta) and make the ecological network transformations consistent. The research group, before starting the laboratory, presented and explained the guidelines and the maps to safeguard and enhance the rural landscape drown in the Strategic Plan of the Franciacorta geographical area.

The co-mapping process to reach shared legends generated two types of results. The first result is associated with the co-production of maps based on shared knowledge of what is the expected information and how it should look like, according to different municipal technicians involved in the participatory process. As a second result, the laboratory identifies the collaboration needed between municipalities when specific cross-border actions are desirable to safeguard the local landscape (Fig. 7).

Fig. 7 Brainstorming on maps during laboratory activities.

The third was on "a shared legend on hydrogeological assessment" and worked on a flood risk map to represent transformation areas foreseen at the local level, flood hazards, flood risk from the Flood Risk Management Plan according to 2007/60/EC directive, and the hydrogeological risk from Hydrogeological System Plan (*Piano di assetto idrogeologico—PAI*). In the first part of the laboratory, a presentation done by the Region Lombardy office of "Land and civil protection" explained inconsistencies detected in geological and seismic maps drown at the local level. This is particularly tricky in bordering municipalities. From the cartographic point of view, several requests for clarification were needed and the members of the laboratory identified desirable information to be added to the map, for instance marking areas included in the Forestry Plan (*Piano di Indirizzo Forestale—PIF*) as it introduces restrictions on land transformation. All the information added is focused on better specifying the mapped data and the attributes to be shown. As a second result, the laboratory strongly raised the awareness of the importance of using the same "shared legend" to integrate the information in Local Emergency Response Plan maps (*Piano di Emergenza Comunale—PEC*). During this laboratory also common building regulations and land equity principles were discussed.

7. Results of the experience

Despite the positive aspect of the memorandum of understanding (MOU) signed by the municipalities of Franciacorta, the survey demonstrated a rather weak framework in terms of effective interest to implement the Strategic Plan. Twelve of the twenty-two Municipalities replied to the questions specifically related to the Strategic Plan and only seven of them demonstrated to be somewhere engaged in the co-planning process. Leadership is absent, no municipality is politically strong enough to lead the territory. There seems to be a relation between the lack of political capacity—also related to the selection of priority activities—and the foreseen inability of municipalities to respond to the obligations imposed by the Strategic Plan. As a result, municipalities have not been able to put the Strategic Plan into effect. This aspect is not only related to the exogenous factors that influence the action arena, but also to the lack of institutional capacity of these municipalities that minimize receptivity at the local level.[23]

Franciacorta municipalities asked for the Strategic Plan to Region Lombardy as a bottom-up need, but there are still evident inter-municipal conflicts and institutional capacity gaps among municipalities that make the Plan inactive without an explicit and consistent top-down political pressure.

This aspect has been strongly highlighted by the research group during the public presentation done to the stakeholder and to the political arena.

Regarding power asymmetries, two (Gussago and Rovato) of the 22 municipalities involved in the questionnaire declared to prefer the traditional forms of planning (general master plan) and a top-down approach in land management instead of the participatory, open decision-making processes and holistic vision foreseen in the Strategic Plan. The main remarkable problem is the passive position assumed by several municipalities that seem to be NOT engaged in any strategic issue and are completely unable to overcome any path dependency and/or turn ordinary management of their territory into a collaborative approach.

The three laboratories have foreseen different levels of attendance by local officials. As participants, they supported the co-mapping construction process in very different ways. The group was strongly engaged in the activity, it enthusiastically discussed and desired to contribute to the shared legend implementation (an example is shown in Fig. 8) by providing suggestions and recommendations. They were interested in understanding all the documents presented, asking for clarification during the presentations done by the researchers in the three laboratories. We can say that for public officials "sharing knowledge means sharing solutions in a very practical and efficient way."

Regarding soil protection and hydrogeological risk assessment, a holistic spatial approach is possible and desirable.

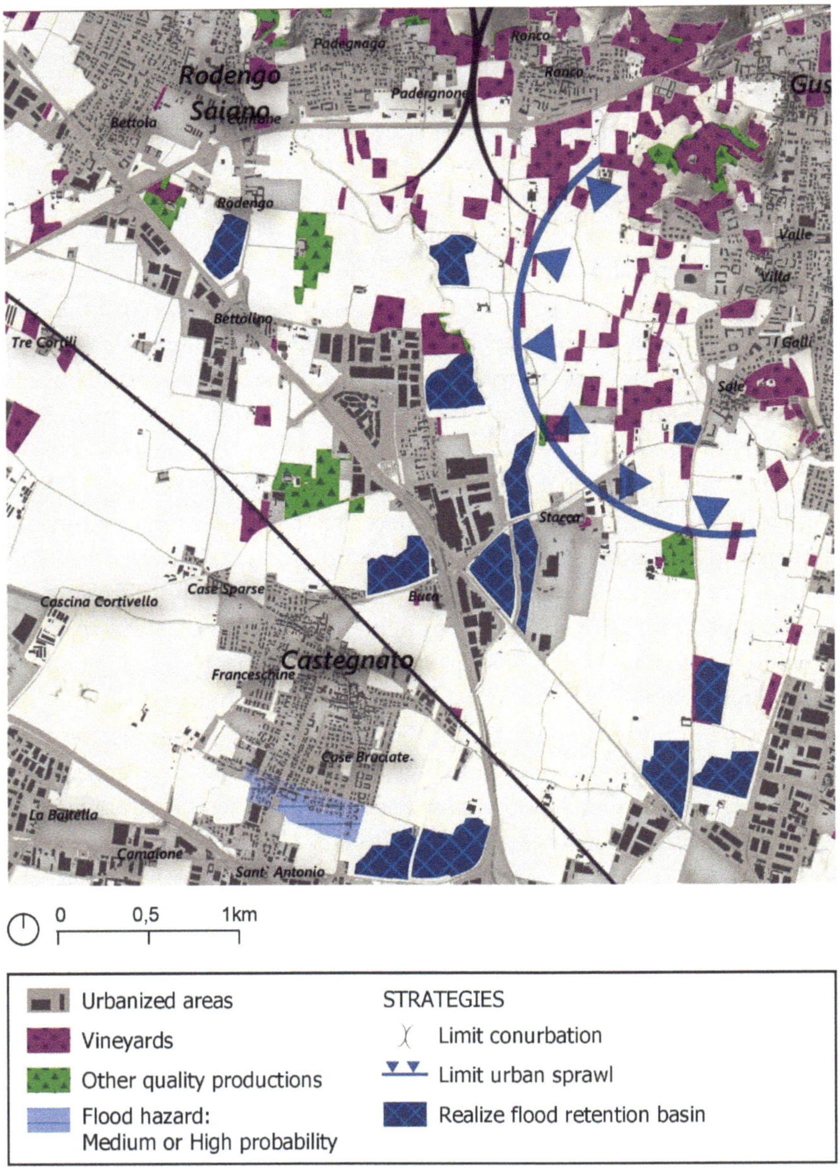

Fig. 8 Strategic Plan of Franciacorta.

There is low awareness of the possibility of a common path between safeguarding the territory from hydrogeological risk and landscape restoration. The sectoral approach is deeply rooted and innovative integrated planning solutions struggle to be accepted. In the case of hydrogeological risk, gray infrastructures are still preferred to green ones. Land equity principles, artificial surface reduction and nature-based solutions as concurring possibilities in taking care of land are solutions looked upon with distrust and perceived as critical issues to be introduced.

8. Conclusions

In Italy there are three main general plans to regulate spatial planning processes: the Regional Plan (*Piano Territoriale Regionale*), which provides a general view of the development of the region; the Provincial Plans (Piani Territoriali di Coordinamento Provinciale), which provide the main objectives related to spatial planning; and the municipal plans (in Lombardy they are called "Piani di Governo del Territorio" according to Region Lombardy Law 12/2005), which constitute the only decision-making level of land transformation. In Italy each Municipality is therefore responsible for drawing up the municipal plan and managing land transformation within the municipality borders. Generally this function is not adequately performed due to lack of continuous updates of the local administrations regarding regulatory, technological, and cultural requirements and innovation that change over the years. Italian Municipalities used to proceed in a scattered order, in some cases renouncing (or not being able) to measure themselves with the actual extent of the transformations taking place, or without being able to overcome the local political visions, an aspect that strongly limits the implementation of strategic visions of supra-municipal level. The result is a mosaic of plans with different objectives and visions, sometimes also in contrast, which insist on a common landscape. In such an heterogeneous context, The Strategic Planning (Area Regional Territorial Plan/*Piano territoriale regionale d'Area*) is also foreseen by Region Lombardy law 12/2005, but it is usually limited to specific infrastructure development needs such as Airports. The Strategic Plan of Franciacorta is an unique example for Italy of a supra-municipal strategic plan specifically focused on an harmonious development of the territory to limit soil sealing and enhance landscape and flood protection.

Indeed, the governance process has brought many benefits: it allowed to create a common system of knowledge of the Franciacorta region, as well as a reading of the territory that goes beyond the municipal administrative boundaries. Besides, the co-design laboratories allowed to build a network of municipal technicians, to highlight shared values, and define shared strategic guidelines. Nevertheless, the experience of Franciacorta is also a story of partial failure: there is considerable uncertainty concerning the effective implementation of the actions foreseen in the Strategic Plan of Franciacorta geographical area 4 years after its approval.

Strong institutions, such as Region Lombardy, play a decisive role in fostering the transformative capacity of a territory, upholding the efforts that municipalities make in their transformative journey. The local capacity to absorb new knowledge (i.e., "transformative capacity") is related to the collective effort made by the different public (e.g., Region Lombardy) and private actors (e.g., the Franciacorta wine consortium) working together. This effort requires a broader scale of societal activities and the formation of appropriate and inclusive governance mechanisms that include all the stakeholders. The focus of change should be the administrative culture as new common values, priorities and routines. A key role can be played by municipal technicians among local officials and their involvement concerning common and innovative land-use issues is essential.

References

1. Mostert E, Junier S. The European flood risk directive: challenges for research. *Hydrol Earth Syst Sci Discuss* 2009;**6**:4961–88. https://doi.org/10.5194/hessd-6-4961-2009.
2. Vitale C, Meijerink S. Flood risk policies in Italy: a longitudinal institutional analysis of continuity and change. *Int J Water Resour Dev* 2021;1–26. https://doi.org/10.1080/07900627.2021.1985972.
3. Vitale C, Meijerink S. Understanding inter-municipal conflict and cooperation on flood risk policies for the metropolitan city of Milan. *Water Altern* 2021;**14**(2):597–618.
4. Tira M, Zazzi M. *Pianificazione territoriale e difesa del suolo. Quarant'anni dopo la relazione «De Marchi»*. Roma: Gangemi Editore; 2013. ISBN:8849223668.
5. Pahl-Wostl C, Becker G, Knieper C, Sendzimir J. How multilevel societal learning processes facilitate transformative change: a comparative case study analysis on flood management. *Ecol Soc* 2013;**18**(4):58. https://doi.org/10.5751/ES-05779-180458.
6. Gallozzi PL, et al. *ReNDiS 2020. La difesa del suolo in vent'anni di monitoraggio ISPRA sugli interventi per la mitigazione del rischio idrogeologico, Rapporto ISPRA n328*; 2020. p. 1–155.
7. ISTAT. *Classificazioni statistiche e dimensione dei comuni al 01/03/22*; 2022.
8. Ruggeri G, Mazzocchi C, Corsi S. Drinking biodiversity: a choice experiment on Franciacorta sparkling wines. *Br Food J* 2020;**122**(8):2531–49. https://doi.org/10.1108/BFJ-06-2019-0451.
9. Regulation (EU) n.d. Regulation (EU) no 1308/2013 of the European Parliament and of the Council of 17 December 2013, establishing a common organisation of the markets in agricultural products and repealing Council Regulations (EEC) No 922/72, (EEC) No 234/79, (EC) No 1037/2001 and (EC) No 1234/2007, https://ec.europa.eu/info/food-farming-fisheries/food-safety-and-quality/certification/quality-labels/geographical-indications-register/details/EUGI00000003568.
10. Richiedei A. Franciacorta: a rich territory, but deprived of its own image. In: Ventura P, Claderon J, Tiboni M, editors. *COST action C27 sustainable development policies for minor deprived urban communities*. Milano: McGraw-Hill; 2011. p. 321–34. ISBN:978-88-386-7282-8.
11. ISTAT. *Censimento della Popolazione al 01/03/2022*; 2022.

12. Ghiglieno I, et al. Characterisation of Franciacorta vineyard agroecosystem to support biodiversity and ecosystem services. *Bio Web Conf* 2022;**44**:1–6. https://doi.org/10.1051/bioconf/20224402003.
13. Bitner MJ. Servicescapes: the impact of physical surroundings on customers and employees. *J Mark* 1992;**56**(2):57–71. https://doi.org/10.1177/002224299205600205.
14. Alampi Sottini V, et al. Winescape perception and big data analysis: an assessment through social media photographs in the chianti classico region. *Wine Econ Policy* 2019;2212-9774. **8**(2):127–40. https://doi.org/10.1016/j.wep.2019.07.001.
15. Carlsen J, Boksberger P. Enhancing consumer value in wine tourism. *J Hosp Tour Res* 2015;**39**(1):132–44. https://doi.org/10.1177/1096348012471379.
16. Karagiannis D, Metaxas T. Sustainable wine tourism development: case studies from the Greek region of Peloponnese. *Sustainability* 2020;**12**:5223. https://doi.org/10.3390/su12125223.
17. Antrop M. Why landscapes of the past are important for the future. *Landsc Urban Plan* 2005;0169-2046. **70**(1–2):21–34. https://doi.org/10.1016/j.landurbplan.2003.10.002.
18. Mirumachi N, Allan JA. Revisiting transboundary water governance: power, conflict cooperation and the political economy. In: *Proceedings of the CAIWA international conference on adaptive and integrated water management: coping with scarcity, Basel, CH*; 2007. p. 1–21.
19. Shao Y, Lange E, Thwaites K. Defining local identity. *Landsc Archit Front* 2017;**5**(2):24–41. https://doi.org/10.15302/J-LAF-20170203.
20. Adobati F, Olivieri A, Pavesi FC, Pezzagno M, Tira M. Franciacorta: un brand (e un piano) per molti paesaggi. In: *AAVV. Cambiamenti. Responsabilità e strumenti per l'urbanistica a servizio del Paese. vol. unico.* Planum Publisher; 2016. p. 181–8. ISBN:9788899237080.
21. Adobati F, Ferlinghetti R, Signoretti F, Lorenzi MA. Dalla Franciacorta al Franciacorta (e ritorno). *Geografie e brand. EyesReg* 2017;2239-3110. **7**(2):54–9.
22. Tira M, Pezzagno M. Prime riflessioni sull'esperienza del Piano Territoriale d'Area della Franciacorta. In: *XXXVII Conferenza Italiana di Scienze Regionali*; 2017. https://www.aisre.it/archivio-abstract-e-papers-dal-2004/?aisre_task=show&out=html&view=list&aisre=1&layout&action&request=1&stato=1&id_presidente=0&filter_search&filter_association_timespan_id=15&filter_author=pezzagno&filter_session_id&filter_abstract_theme_id&filter_keywords.
23. Butler M. Managing from the inside out: drawing on 'receptivity' to explain variation in strategy implementation. *Br J Manag* 2003;**14**(Special Issue):S47–60. Blackwell Publishing Ltd.

CHAPTER FIVE

Soil quality and urban sprawl: Insights from long-term patterns in the Rome metropolitan region

Francesca Perrone[a,*], Francesco Gallucci[b], Enrico Maria Mosconi[c], and Luca Salvati[d]

[a]Department of Planning, Design, and Technology of Architecture (PDTA), Faculty of Architecture, Sapienza University of Rome, Rome, Italy
[b]Consiglio per la Ricerca in agricoltura e l'analisi dell'Economia Agraria (CREA), Monterotondo, Italy
[c]Department of Economics, Engineering, Society and Business Organization (DEIM), University of Tuscia, Viterbo, Italy
[d]Department of Methods and Models for Economics, Territory and Finance, Faculty of Economics (MEMOTEF), Sapienza University of Rome, Rome, Italy
*Corresponding author: e-mail address: francesca.perrone@uniroma1.it

Contents

Abstract

The dynamics related to land take depend on the interaction of two factors: soil quality and anthropogenic actions and strategies. Soil as "living system" plays an active role in the phases of interaction between the components of ecosystems. Meanwhile, the soil is exposed to intense and constant human-induced degradation processes. The aim of this paper is to quantify the impact of urban expansion on soil quality in a Mediterranean urban area (Rome, Italy). Over an examined area of 1500 km^2, urban areas increased by 5.9% *per year* (from 8.2% in 1949 to 36.6% in 2018). At that time, urban areas consumed high-quality soils in greater proportion compared to low-quality soils while croplands and forests progressively covered low-quality and partially degraded soils. Moreover, dispersed peri-urban settlements have been built in Rome on high-quality soils rather than in dense urban settlements, suggesting that the recent low-density urban expansion mainly affects soil quality and land resources. Those data show that sprawl consumes high-quality land at higher rate than compact growth, thus

Advances in Chemical Pollution, Environmental Management and Protection, Volume 8
ISSN 2468-9289
https://doi.org/10.1016/bs.apmp.2022.10.012

influencing the environmental quality of neighboring land. This case study, as an assessment and monitoring of soil quality over multiple timescales, is the basis on which any planning project can be built. It develops, in a scenario where, although significant progress in the analyses and assessments of land take has been achieved, it still appears as "open challenge" and it is not easy to work on a more effective decrease in land take and a consequent conservation of ecosystem functions and services.

Keywords: Land-use changes, Soil degradation, Urban expansion, Land take, Mediterranean urban region, Spatial planning

1. Introduction

"Soil quality is one of the three components of environmental quality, besides water and air quality."[1] Soil plays multiple indispensable functions,[2–4] as a "resource"[5] to maintain the ecological-environmental balance and support biological productivity[6] as well as a "natural capital"[7,8] to ensure the survival and well-being of man.[9]

Soil constitutes the main physical medium for Ecosystem Services (ES).[3,10] "Soils result from complex actions and interactions of processes in time and space and hence are themselves diverse in form and properties and the level of ecosystems services they provide" (Ref. 11: art. 4). The soil allows life on land; is the main biomass container on the planet; acts as guardian of biodiversity; serves as a "basis" for all human activities; and underlies the production of food, raw material and some components of medicines (Provisioning services). Additionally, soil represents one of the main natural carbon sinks, second after the oceans; and is a regulator of water flows and influences the quality of the water (Regulating services). Furthermore, soil constitutes an essential element for the preservation of cultural heritage (historical-archeological and ecological-environmental) (Cultural services).[10,12,13] Because of the relevant services provided, the soil must be considered an indispensable resource.[14,15]

Yet, the soil is highly threatened.[3,16] It is exposed to a loss and/or reduction of the main properties of a physical, chemical and biological nature.[17] This occurs due to: erosive processes, overflows of water, increase of soil salinity, drop in microorganisms, soil organic matter and biological productivity decline, global warming, increase in urban heat island, decline of soil fertile, and loss of natural and seminatural landscapes.[18–21] In addition, there are the effects of soil compaction and soil sealing.[22] These degradation processes represent a variety of contributing factors which, acting simultaneously, make the soil a fragile and easily alterable resource. To confirm this, we should just consider that "[...] 75% of the lands above sea level suffer deterioration due to the presence of man."[23]

The anthropogenic activities themselves, if on the one hand, lead to the conditions for balance in relations to change between the environmental and anthropogenic system, on the other hand, they negatively affect the equilibrium through actions triggering soil degradation. Indeed, the uncontrolled and growing urbanization, infrastructure, industrialization and contamination of the territory, is not followed by an appropriate planning and design of interventions, consistent with the criteria of environmental, social and economic sustainability.[24] Massive urbanization mainly driven by urban sprawl[25] and informal growth, peri-urbanization, small towns and planning deregulation have caused landscape changes, environmental degradation, consequences on a social, economic and environmental level, as regards, for example, the impact on ecosystem services[26,27] and urban sustainability.[28–30] It is following by continuous obstacles in partition of land-use planning decision making, for example, concerning the tight relation between built-up areas and fertile lands bordering existing urban areas.[31]

Numerous papers explored the ecological consequences of the urban expansion in terms of soil resources and landscape fragmentation[32] with a special focus on point and diffused processes of land/soil degradation and land take.[33–37] Contrary, restricted information are available on physical and biological degradation of urban and peri-urban soils.[38] Soil resource depletion due to urban growth was poorly monitored at both regional and local scales.[39–41] Yet, Prestele et al. in van Vliet et al.[31] "shows that five out of eleven global-scale land change projections represent built-up area, while the other six include only natural and agricultural land."

In Europe, due to land take, portions of natural soil are lost every year, for a territorial extension equal to the size of a city like Berlin. In southern Europe, an ecologically-fragile region due to soil aridity, climate change, and diffused anthropogenic pressure, urbanization caused habitat fragmentation, agricultural land abandonment, and loss in biodiversity, indirectly affecting soil quality and land degradation.[42–46] While Mediterranean cities have been traditionally described as compact and hyper-dense,[47] since the early 1990s urban sprawl has been increasingly observed in these cities.[19,39,48–58]

Containing the processes of uncontrolled spread of a city and monitoring "the interrelations between land-use/cover changes and environmental impacts of urbanization"[59] means safeguarding the balance between human intervention and resilience of natural systems,[60] with particular attention paid to the idea of sustainability, both urban and territorial, and to the resulting planning and design practices. Decision-makers must pay attention to the risk factors determined by deregulated urbanization and consider the protection of "highly qualified" soils as the main objective.[61]

Unfortunately, policies promoting urban containment have been poorly effective in planning-deregulated Mediterranean regions, determining an increase in the risk of desertification at local scale.[62] It is also why the sensitivity of land degradation requires a standing monitoring and a more efficient coordination of information, to hypothesize increasingly effective interventions.

Based on a diachronic analysis (1949–2018) of land-use changes and soil characteristics in a southern European urban region (Rome, Italy), the present study verifies if (i) long-term growing urbanization takes up areas with different soil quality compared to agricultural and forest land, and if (ii) dispersed, low-density settlements consume land with higher soil quality than compact and dense settlements. In other words, this study tests if landscape transformations due to urban expansion are reflected in changing soil quality and if changes in urbanization patterns may reflect different rates of land take.

2. Methods

2.1 Study area

Rome is one of the largest Mediterranean urban regions, located in Lazio region (41° 53′ 35″ N; 12° 28′ 58″ E). The hinterland of the city consists of other 121 municipalities. The updated estimation of 2021 indicates a total number of inhabitants equal to 4.2 million, distributed over a territorial area of 5363 km^2 and a population density of 788.26 inhabitants/km^2.[63] The orography of the study area is characterized by three main mountain systems: Lazio Anti-Apennines; Lazio Sub-Apennines; and Alban Hills and northern side of Monti Lepini. The coast is border with the Tyrrhenian Sea, for about 133 km. The main rivers that cross the city of Rome are the Tiber, followed by the Aniene, and there are some lakes, including Bracciano, Albano and Nemi. Moreover, the Mediterranean climate characterizes the western coast, while the attenuated continental climate characterizes the eastern inland areas.

In particular, the investigated area encompasses the boundaries of the municipalities of Rome and Fiumicino (1500 km^2). This region is characterized by a complex topography consisting of 70% lowlands and 30% uplands.

Although urban areas occupy an important (and increasing) part of the region, the majority of the prefecture's area still consists of forests, pastures, and cropland (Fig. 1).

2.2 Data

Land-use data were obtained from the elaboration of two maps available for the same area: (i) the 1:25,000 land-use map produced by Italian Military Geographical Institute (IGM) in 1949 and (ii) the Urban Atlas 1:10,000 land-use digital map produced by the European Environmental Agency in 2018 (see also[64]). The 1949 map was derived from interpretation and vectorization of a set of Istituto Geografico Militare (IGM) 1:25,000 digital topographic maps (the reference system is UTM33 ED50) carried out within a Geographic Information System environment (ArcGis 9.2, Esri Inc., Redwoords, USA). To minimize errors, the same person carried out interpretation and delineation of land-use polygons at each map. We identified polygons of three different land-use types (urban areas, agricultural and pasture land, forests) on the IGM raster maps with the help of a zoom stereoscope and expert knowledge of the territory based on field surveys, old photos, and literature data.[65]

The 2018 Urban Atlas map include 19 land-use classes that were mainly based on Corine Land Cover nomenclature. These were aggregated into three representative categories (urban areas, agricultural areas and pastures, forests) according to the available classes in the 1949 IGM map (for further information, see Urban Atlas). Urban Atlas is the major initiative dealing with low-resolution land-use monitoring in Europe and represents an information tool for both assessment and policy analysis.[66] The Urban Atlas project was undertaken by the European Environmental Agency within the Global Monitoring for Environment and Security (GMES) framework. The Urban Atlas service offers a high-resolution digital land-use map of several European urban areas. More than 700 cities in the European Union were covered in 2018 (for further information, see Urban Atlas).[67]

A 1:50,000 digital soil map produced in the early 198 and covering the municipalities of Rome and Fiumicino was finally used in the present study as a reliable pedological reference for the area. This map classified the different soil types found in the area according to a field survey and a Digital Elevation Model as ancillary data source, evidencing selected limiting factors (erosion, compaction, pollution). These degradation processes are all-important "environmental factors that—by its decrease, increase,

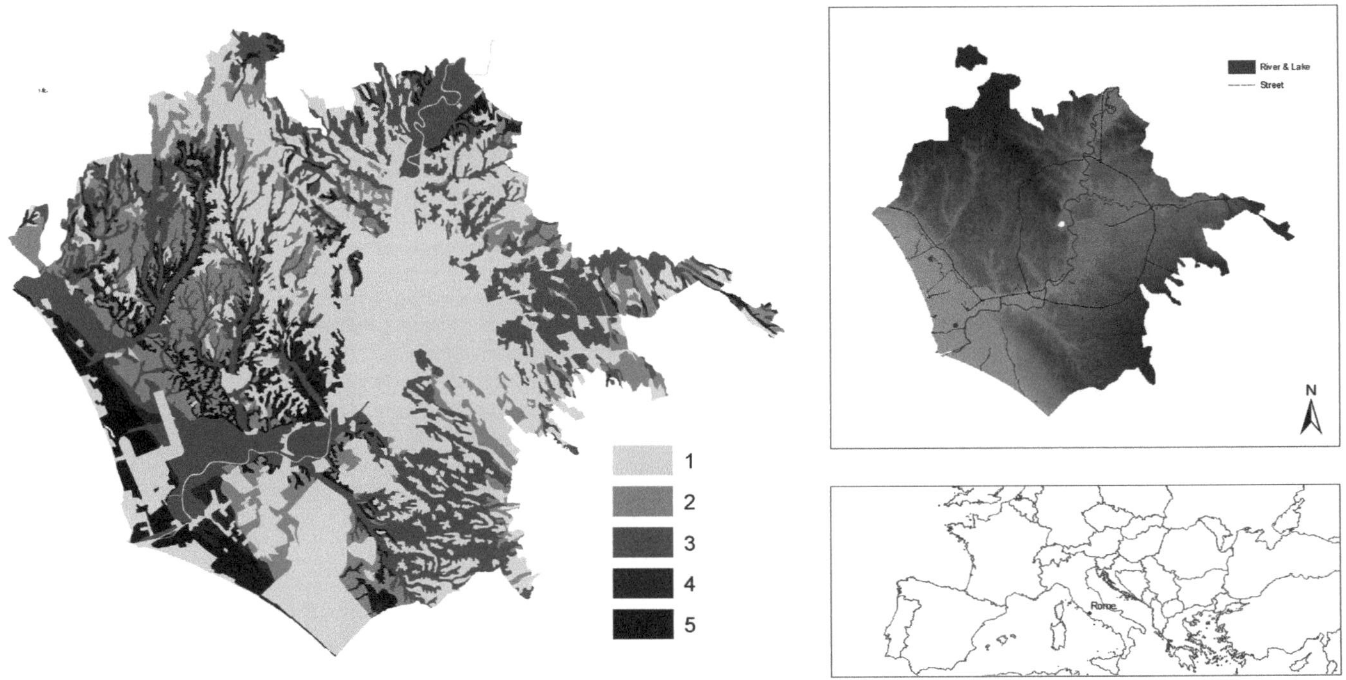

Fig. 1 Maps of the study area illustrating (right) elevation, main roads and rivers, and (left) the soil capability index using five classes (from 1. croplands without limitations to 5. croplands with relevant limitations).

absence, or presence—limit the growth, metabolic processes, or distribution of organisms or populations" (Oxford Reference). Soil erosion—triggered by different factors (such as intrinsic features, steep slopes, inappropriate land use, ecological disasters, etc.)—causes the loss of soil functions and the loss of soil itself. "The Mediterranean region is historically the most severely affected by erosion" (Ref. 68; for further information, see also[69]). Soil compaction—triggered by mechanical pressures—causes the reduction of the pore space between soil particles and the soil loses, for example, its water storage capacity and stability. It too can be considered an EU-wide problem.[68] Diffuse pollution—triggered by atmospheric deposition, certain farming practices and inadequate waste and wastewater recycling and treatment—causes the reduction of the buffering capacity of soils and in addition soil acidification, loss of soil fertility and damage of biological activity. "The cost of diffuse soil contamination is not seen so much in soil itself, but in the consequences of the breakdown of the buffering capacity of soil" (Ref. 68; for further information, see also[69]). The soil map of Rome was referenced in UTM33 ED50 system. A part of the publication with all the technical details referring to soil types and cartographic assessment can be downloaded from a public website held by Italian Institute for Environmental Protection and Research.[70]

The Municipality of Rome (Dpt. X—II° U.O. Sustainable Development) considered the publication of the "Soil Map" and the information data to be of great usefulness, in order to (i) promote knowledge of the characteristics and peculiarities (historical-monumental and ecological-environmental) of the Roman territory and (ii) encourage research and activities for the purpose of allowing a sustainable use of the territory and "a design and planning approach to safeguard and enhance environmental resources".[71] The study also provided an important contribution to the programmatic guidelines of the Rome PRG, as regards, for example, the environmental system and the definition of the "Ecological network." The graphic and data processings are related in particular to: "the characteristics of the soils," "the degree of drainage," "the capacity to use the soils," "the level of environmental risk (level of pollution risk)," "the possibilities and value of the agro-environmental use of soils." The main types of soil degradation in Rome taken into account are: erosion, compaction and pollution (limiting factors illustrated above), in addition to the spreading of urban developments and salinization. According to Arnoldus-Huyzendveld[71] "in the territory of Rome, the risk of pollution is present everywhere, but the intensity strongly depends on the proximity of the sources and the type of agricultural use. The degradation by soil compaction occurs in the Roman territory in the arable lands on large surfaces, in particular on soils with a fine grain size and with the aquifer close to the surface." The risk of erosion is lower than in other areas of Italy, but the evaluated values are worrying anyway, for example, for loss of soil organic matter, eutrophication, formation of deep soil incisions, etc. (for further and better particulars, see[71]).

2.3 Assessment of soil quality depletion due to urbanization

Based on the spatial distribution of limiting factors (see above), soil map was overlaid to the land-use maps to derive information on high- and low-quality soil consumption caused by urbanization in Rome. Soil quality depends on "the capacity of a specific kind of soil to function, within natural or managed ecosystem boundaries, to sustain plant and animal productivity, maintain or enhance water and air quality, and support human health and habitation" (Natural Resources Conservation Service—USDA). The surface area with defined soil attributes (according to the factors previously described) within each land-use class was determined in 1949 and 2018 using three land-use classes, and in 2018 considering 18 land-use classes by the intersection of land-use polygons with the soil map polygons using the ArcGIS "intersect" tool. Based on this rationale, we calculated (and illustrated in a tabular format) the percent soil area (in total landscape area) classified according to selected limiting factors in Rome ("compaction," "erosion," and "pollution"). These degradation processes were considered the most frequent in the study area. We evaluated also the surface area of soils actually unaffected by degradation processes.[72–74]

3. Results

The present study verifies that in Rome urban settlements covered 123 km^2 in 1949 (8.2%) while agricultural and forest land covered respectively 1184 km^2 (79.0%) and 169 km^2 (11.3%). Between 1949 and 2018 urban areas increased by 5.9% *per year* extending over 549 km^2 (36.6%) in 2018, while agricultural and forested lands decreased to 835 km^2 (55.7%) and 135 km^2 (9.0%), respectively. In 2018, compact and dispersed urban fabric occupied 25.9% and 18.9% of the impervious surface, respectively (Figs. 2 and 3).

Compared to agricultural areas and forested land, urban areas occupied preferentially soils with high land ability during the last 70 years (Table 1).

Fig. 2 Comparison between two satellite images from 1943 (left) and 2018 (right), showing the evolution and expansion of the urban settlement in the Zone of Rome Val Melaina, traced within the area of the Ager Romanus. According to the toponymic subdivision is indicated by the initials Z. I. (A). According to Urban subdivision it is identified by Urban zones of the Municipio Roma III of Rome Capital (Administrative subdivision) (B). *Source: Di Sannita—Opera propria, CC BY-SA 4.0; Google Earth. https://commons.wikimedia.org/w/index.php?curid=90839419*

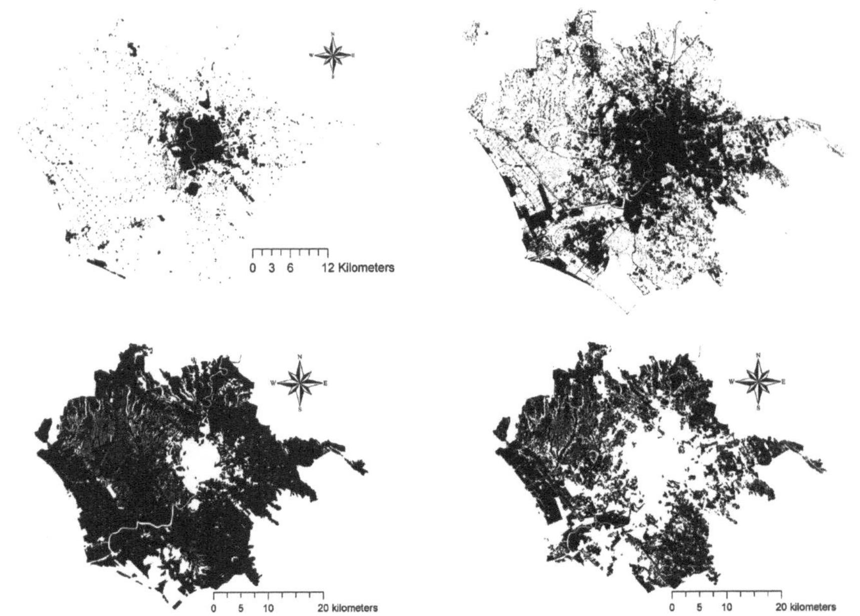

Fig. 3 Land-use in Rome in 1949 (left) and 2018 (right); upper line: urban settlements; lower line: agricultural areas.

Table 1 Percent soil area classified according to selected limiting factors in Rome by basic land-use and year.

	1949			**2018**		
Class	**Urban settlements**	**Cropland**	**Forests**	**Urban settlements**	**Cropland**	**Forests**
Absent	21.2	21.5	12.0	23.2	20.4	9.7
Compaction	0.6	1.8	1.9	1.0	1.9	2.8
Erosion	17.4	13.8	25.1	11.0	15.6	26.3
Pollution	9.6	8.3	1.5	10.9	7.3	1.6

Soils with no limiting factors (such as erosion, compaction, pollution) were mainly occupied by the shrinking urban areas (+2%) while agricultural and forest land expanded on soils with important production limitations due to erosion (+1.7% and +1.2%, respectively). Interestingly, urban areas expanded also on soil affected by diffused pollution and/or point contamination (Table 2). Considering a high-resolution nomenclature for 2018 (Table 2), low-density residential settlements expanded preferentially over

Table 2 Percent land classified according to land capability and sustainability in soil use in Rome by detailed land-use class in 2018.

Soil variables	Continuous urban fabric (S.L. > 80%)	Discontinuous dense Urban fabric (S.L.: 50–80%)	Discontinuous medium Density urban fabric (S.L.: 30–50%)	Discontinuous low density Urban fabric (S.L.: 10–30%)	Discontinuous very low density Urban fabric (S.L. < 10%)	Isolated structures	Industrial, commercial, public, military and private units	Fast transit roads and associated land	Other roads and associated land	Railways and associated land	Airports	Mineral extraction and dump sites	Construction sites	Land without current use	Green urban areas	Sports and leisure facilities	Agricultural + Semi-natural areas + Wetlands	Forests
Absent	14.3	21.6	26.0	34.6	40.9	26.5	21.9	17.5	22.7	21.0	0.3	22.6	22.1	15.4	18.5	22.0	20.4	9.7
Compaction	0.0	0.8	0.2	0.0	0.0	1.1	0.6	2.3	1.0	0.1	0.0	11.0	0.0	1.5	0.0	1.8	1.9	2.8
Erosion	7.3	8.1	10.8	14.8	26.5	17.6	8.2	10.4	11.4	6.2	0.0	8.4	15.4	6.8	17.3	10.8	15.6	26.3
Pollution	23.3	16.9	8.9	6.5	5.1	5.2	14.1	8.8	10.3	8.2	0.0	8.2	6.9	7.7	8.9	7.9	7.3	1.6

soils without limiting factors while compact urban fabric occupied soils affected by pollution limitations more frequently than dispersed settlements. However, dispersed settlements consumed also more land susceptible to soil erosion than compact settlements.

4. Discussion

In Italy, in the last few years, a territorial fragmentation of about 30% of the total surface, as well as a loss of 28% of utilized agricultural land, in addition to a substantial increase in carbon dioxide in the atmosphere and an increase of about 260 million liters of more water to manage, has occurred.[15] The information specified in the report on "Consumo di suolo, dinamiche territoriali e servizi ecosistemici,"[75] shows that, between 2020 and 2021, there was an increase of 0.3% in man-made surfaces (artificial roofs and interventions), at national level, equal to 69.1 km^2, for a total of 21,485 km^2 (86% of which located on useful soil), that is 7.13% of the national territory.

Furthermore, we have to consider that the "degree of urbanization evaluated through the density of artificial surfaces" continues to increase at the national level: high-density urban areas occupy 2.9% of the territory, equal to 8768 km^2 and medium/low density suburban areas cover 15.4% of the territory, accounting for 46,453 km^2.[75]

Therefore, the general trend of the problem, on a national level, turns out to be quite critical, because, if up to the 195, at the regional level, the percentage of land take did not, in any case, exceed 5% of the reference territory, whereas in 2015 it was found that, for most of the Regions of Italy, the percentage of land take was equal or greater than 7%.[76] In 2021, the Region of Lazio currently stands at between 7 and 9%, with an increase in land take of 4.07 Km^2, between 2020 and 2021. Furthermore, as for the degree of urbanization at the regional level, the extension of high-density urban areas is 548 km^2, while low/medium density suburban areas occupy 3340 km^2.[75] According to Munafò[75] both the Metropolitan City of Rome Capital, reaching about 701 Km^2 to 2021 (in addition to the 2.16 km^2 of 2020 and 2021, of which over 0.95 km^2 in the municipal area of Rome), and Rome as the Capital of Italy, exceeding 303 km^2 to 2021 (in addition to the 0.95 km^2 of 2020 and 2021), rank first in terms of net increase in land take.

This paper provides an original contribution to the issue of natural resource depletion caused by urban expansion by illustrating a methodology to evaluate the quality of soil consumed by city's growth and the impact of different development paths on soil quality at regional scale. A large Mediterranean urban region characterized by ecological fragility and susceptibility to land degradation was selected as a case study. The illustrated approach is implemented on a Geographic Information System interface using diachronic land-use cartography and a high-resolution soil map. Results of the present study demonstrate that urban expansion consumes preferentially soils with high land capability, high potential productivity for cropland, and low susceptibility to soil deterioration (see also[54,77–80]). As a consequence, cropland, pastures, and forests have progressively occupied lower quality soils with some degradation problems (e.g., higher susceptibility to soil erosion and contamination). In the long-term, this process may have an impact on landscape resilience and even ecosystem stability. This is especially true in regions undergoing climate changes and increasing population.[81] Although soil sealing is seen as a major soil deterioration problem in Europe, and therefore monitored at both national and local scales,[44] diachronic approaches quantifying soil resource depletion due to urbanization, like the one illustrated in this paper, are especially needed to inform sustainable regional planning. Long-term analysis and monitoring approaches have the task of informing policies for the conservation of high-quality soils in urban/peri-urban, natural/semi-natural areas and studies that explore the complexity of soil-landscape relationships in contexts heavily urbanized. It follows that it is absolutely essential to know, understand, interpret and monitor the ongoing territorial phenomena, in order to design and draw up effective tools, capable of administering, coordinating, managing and resolving some "territorial issues," as with land take processes as well as its resources deterioration.[82]

The study also verified that over the past 70 years the urban sprawl development phase consumes high-quality soils in a significantly higher proportion than the compact growth phase. These results are in agreement with previous studies carried out in the same socioeconomic context and demonstrating how urban diffusion (i) consumes more natural and semi-natural land, (ii) impacts land uses with higher vegetation quality, and (iii) fragments a higher proportion of agricultural and forest land compared to dense urban growth.[64] Based on these results, the diffused, low-density urban form in the Mediterranean basin can be seen as less sustainable than the traditional compact city from the land take point of view.

In regions experiencing planning-deregulated urbanization, exurban development reflects the transformation of the peri-urban land into a rural-non-farm landscape with implications for the quality of soils and their possible degradation.[83,84] Besides the monitoring of damage caused by land take, the ability to control and manage the problem looks extremely complicated. Measures designed to cope with the future development paths for large Mediterranean cities and capable to promote a "compact polycentric" development may contribute to determine land-saving urban forms (such as the neotraditional development, the urban containment, the compact city, and the eco-city; see[85] for further information) also effective in soil conservation.[86–88]

On the basis of the above, to result completely effective, conservation of traditional landscapes and vegetation associations in the northern Mediterranean region should integrate measures for the containment of land take and the promotion of sustainable urban forms. Concerning the municipality of Rome, the city plan enforced in Rome (Piano Regolatore Generale del Comune di Roma—PRG 2008) based on some "cardinal rules," including "the principle of sustainability" and "the priority of redeveloping the existing building stock" and introduced some conservation measures for natural and semi-natural lands surrounding the urban area (for further and better particulars, see "Rete ecologica," prescriptive graphic art of PRG 2008). In the general Report of the Plan, among "the connotations of the urban plan," we refer to the importance of the "redevelopment of the city." It is stated that "assuming the environmental and historical value of the still unused space, a settlement model has been built, that starts from the identification and protection of free spaces and delimits the space within which it is possible to intervene for building transformations. The size of the new building is a minimum percentage of the size of the city (about 64 million cubic meters, of which 41 have already been planned, compared to the approximately 702 million cubic meters of the built city, therefore equal to 9% increase in overall buildability and to 3% of buildability to be programmed). The location of these volumes is in the central areas (20.9%) and mainly concerns non-residential destinations. [...] As for the soil resource, the plan guarantees, first of all, the interruption of extra-urban land consumption. The final quantity of free areas envisaged in the new plan is approximately 87,740 hectares" (Ref. 89; for further and better particulars, see "Report," descriptive report of PRG 2008). At the same time "the measures, actions, interventions in the ecological network are aimed at preserving, enhancing, restoring the values and levels of naturalness of the areas, as well as ensuring their integration according to criteria and objectives of geographical continuity and ecological functionality."[90]

Nevertheless the "effectiveness" in sprawl containment and land take is rather unclear.[52] Land consumption monitoring actions try to counter the intensification of threats, but they often fail, and in the meantime, soil degradation moves forward, as we have already seen, at an ever quicker and not sustainable pace.

Basically, there is a lack of available data, which can be considered "complete," relating to the soil consumption and the set of degradation processes that determine the alteration of soil and the consequent losses, as in the case of the dimensions of density, continuity and fragmentation of urban sprawl. Consequently, it happens that the operational tools (such as city plans and spatial planning systems) develop approaches which, in the end, prove to be ineffective at the level of urban and territorial planning interventions. Those tools, at the same time, lose sight of a further objective that should be taken into consideration, namely the identification and provision of programmatic actions for the monitoring of land consumption and its main dynamics that determine its evolution and characteristics.

The balance between the process of artificialization of the soils and the resilience capacity of the environmental system (in its entirety) is related to the sustainability of the planning and monitoring tools of territorial resources and, consequently, their capacity to direct the decisions and ensure efficient performance as regards land use.[91]

The major problem is the "inability of man" to imagine and introduce sustainability in interventions for the artificialization of soils, due to the effect of several "coefficients": methods of approach; implementation of proceedings; choice and use of materials, etc. Unlike agricultural practices, which are usually tolerable based on soil conditions, the whole of the practices related to urbanization or other forms of artificialization of soils are in the majority of cases unsustainable for soils. To confirm this, we should just consider that according to Marchetti et al.[23] "land take means, firstly, the distortion of the landscape, a fracture in the consolidated equilibrium between the city and the countryside, new environmental expenses in terms of the use of natural resources and waste disposal, the redefinition of social identities."

5. Concluding remarks

The soil, as an ever-changing "living body," provides ES, as the final product of the "ecosystemic supply chain" in favor of mankind (in terms of social value, yield and economic savings, long-term healthiness, etc.). Soil, as a slowly evolving "interacting system," consequently depends on the physical-structural, environmental and climatic conditions of the context, which are the cause of the alternation and frequency of natural disasters. It also depends on the intensity of the anthropogenic action, the type and quality of the interventions, which lead to alterations and losses in the functions of soil, such as production, support, purification and filtering.[82] The idea underlying this research is the hypothesis that the diachronic analysis (1949–2018) of the impact of urban expansion on different soil quality in a Mediterranean urban region (Rome, Italy) could represent an additional technical and fact-finding element, and that it is able to exert significant influence on the growth of both widespread awareness and technical skills and responsibilities regarding the value and relevance of the "soil element." Soil quality should be incorporated in urban planning strategies as a crucial target when choosing to protect or develop an urban area. This involves planning actions and interventions aimed at simultaneously, monitoring issues relating to: the safeguarding and enhancement of ecological-environmental components; the respect for social needs; and the efficient distribution of economic resources. We cannot sacrifice soils with high potential qualities in favor of urbanization. However, it does not mean sacrificing socioeconomic development and technological progress, but it is necessary to understand the limits of the approach adopted, in order to optimize it. In this regard, the results introduced in the research study offer room for further discussions that describe different scenarios and development perspectives for planning tools, since they relate: (i) the systematic nature of the recognition and evaluation process of soil quality; (ii) the reliability of the monitoring processes relating to land use changes; (iii) the inclusiveness of urban and territorial developments and uncertainties. This case study gives the occasion to reflect on how to deal with the degradation factors to which soil is consistently subjected, through coordinated actions between the provision of analytical data and monitoring tools, as well as the land use planning that prove to be potentially effective.

References

1. Bünemann EK, Bongiorno G, Bai Z, Creamer RE, De Deyn G, de Goede R, et al. Soil quality—a critical review. *Soil Biol Biochem* 2018;**120**:105–25.
2. Bouma J, van Ittersum MK, Stoorvogel JJ, Batjes NH, Droogers P, Pulleman MM. Soil capability: exploring the functional potentials of soils. In: Field DJ, Morgan CLS, McBratney AB, editors. *Global soil security. Progress in soil science.* Cham, Switzerland: Springer; 2017. p. 27–44.
3. Drobnik T, Greiner L, Keller A, Grêt-Regamey A. Soil quality indicators—from soil functions to ecosystem services. *Ecol Indic* 2018;**94**(1):151–69.

4. Karlen DL, Ditzler CA, Andrews SS. Soil quality: why and how? *Geoderma* 2003;**114**:145–56.
5. Koch A, McBratney A, Adams M, Field D, Hill R, Crawford J, et al. Soil security: solving the global soil crisis. *Global Pol* 2013;**4**:434–41.
6. Bristow KL, Marchant SM, Deurer M, Clothier BE. Enhancing the ecological infrastructure of soils. In: *1h World Congress of Soil Science, Soil Solutions for a Changing World, 1–6 August 2010. Brisbane, Australia*; 2010. p. 1–6.
7. Dominati EJ. Natural capital and ecosystem services of soils. In: Dymond JR, editor. *Ecosystem services in New Zealand—conditions and trends.* Lincoln, New Zealand: Manaaki Whenua Press; 2013.
8. Dominati EJ, Patterson M, Mackay A. A framework for classifying and quantifying the natural capital and ecosystem services of soils. *Ecol Econ* 2010;**69**:1858–68.
9. Robinson DA, Hockley N, Dominati E, Lebron I, Scow KM, Reynolds B, et al. Natural capital, ecosystem services, and soil change: why soil science must embrace an ecosystems approach. *Vadose Zone J* 2012;**11**(1):1–6.
10. Breure AM, De Deyn GB, Dominati E, Eglin T, Hedlund K, Van Orshoven J, et al. Ecosystem services: a useful concept for soil policy making! *Curr Opin Environ Sustain* 2012;**4**(5):578–85.
11. Food and Agriculture Organization of the United Nations—FAO. *Revised world soil charter, environmental policy and law.* Rome, Italy: Food and Agriculture Organization of the United Nations; 2015.
12. Adhikari K, Hartemink AE. Linking soils to ecosystem services—a global review. *Geoderma* 2016;**262**:101–11.
13. Ferrarini A, Bini C, Amaducci S. Soil and ecosystem services: current knowledge and evidences from Italian case studies. *Appl Soil Ecol* 2018;**123**:693–8.
14. Amato M, Di Gennaro A, Mangoni F, Montanarella L, Terribile F. *Sotto i nostri passi: conoscere e comunicare il suolo.* Auletta-Pertosa, Italy: Fondazione Musei Integrati dell'Ambiente (MIdA); 2015.
15. Pileri P. *Che cosa c'è sotto? Il suolo, i suoi segreti, le ragioni per difenderlo.* Milan, Italy: Altreconomia; 2016.
16. Casa M, Pileri P. *Il suolo sopra tutto. Cercasi "terreno comune": dialogo tra un sindaco e un urbanista.* Milano, Italy: Altra Economia; 2017.
17. Gardi C, Jeffery S, Saltelli A. An estimate of potential threats levels to soil biodiversity in EU. *Glob Chang Biol* 2013;**19**(5):1538–48.
18. Glæsner N, Helming K, De Vries W. Do current European policies prevent soil threats and support soil functions? *Sustainability* 2014;**6**(12):9538–63.
19. Munafò M. *Consumo di suolo, dinamiche territoriali e servizi ecosistemici, 2020.* Report SNPA 15/20. Rome, Italy: ISPRA; 2020.
20. Palm C, Sanchez P, Ahamed S, Awiti A. Soils: a contemporary perspective. *Annu Rev Env Resour* 2007;**32**:99–129.
21. Recanatesi F, Clemente M, Grigoriadis E, et al. A fifty-year sustainability assessment of Italian agro-forest districts. *Sustainability* 2016;**8**:32.
22. Artmann M. Assessment of soil sealing management responses, strategies, and targets toward ecologically sustainable urban land use management. *Ambio* 2014;**43**(4):530–41.
23. Marchetti M, Lasserre B, Pazzagli R, Sallustio L. Rural areas and urbanization: analysis of a change. In: *Scienze del territorio.* vol. 2. Firenze University Press; 2014. p. 249–58.
24. Allen A. Environmental planning and management of the peri-urban interface: perspectives on an emerging field. *Environ Urban* 2003;**15**(1):135–47.
25. Hermant-de Callataÿ C, Svanfeld C. *Cities of tomorrow. Challenges, visions, ways forward.* Luxembourg city, Luxembourg: European Commission; 2011.
26. Alberti M. The effects of urban patterns on ecosystem function. *Int Reg Sci Rev* 2005;**28** (2):168–92.

27. Alberti M. Maintaining ecological integrity and sustaining ecosystem function in urban areas. *Curr Opin Environ Sustain* 2010;**2**:178–84.
28. Festus IA, Omoboye IF, Andrew OB. Urban sprawl: environmental consequence of rapid urban expansion. *Malays J Soc Sci Humanit* 2020;**5**(6):110–8.
29. Yasin MY, Yusoff MM, Abdullah J, Noor NM. Is urban sprawl a threat to sustainable development? A review of characteristics and consequences. *Geografia* 2020;**16** (4):56–68.
30. Zipperer WC, Northrop R, Andreu M. Urban development and environmental degradation. In: *Oxford research encyclopedia of environmental Science*; 2020, 1–25. p. 1–25.
31. van Vliet A, Eitelberg DA, Verburg PH. A global analysis of land take in cropland areas and production displacement from urbanization. *Glob Environ Chang* 2017;**43**:107–15.
32. Fiorini L, Zullo F, Marucci A, Romano B. Land take and landscape loss: effect of uncontrolled urbanization in southern Italy. *J Urban Manag* 2019;**8**(1):42–56.
33. Long H, Liu Y, Hou X, Li T, Li Y. Effects of land use transitions due to rapid urbanization on ecosystem services: implications for urban planning in the new developing area of China. *Habitat Int* 2014;**44**:536–44.
34. McKinney ML. Effects of urbanization on species richness: a review of plants and animals. *Urban Ecosyst* 2008;**11**:161–76.
35. Prestele R, Alexander P, Rounsevell MDA, Arneth A, Calvin K, Doelman J, et al. Hotspots of uncertainty in land-use and land-cover change projections: a global-scale model comparison. *Glob Chang Biol* 2016;**22**:3967–83.
36. Schneider A, Logan KE, Kucharik CJ. Impacts of urbanization on ecosystem goods and services in the U.S. corn belt. *Ecosystems* 2012;**15**:519–41.
37. Wang M, Sun X. Potential impact of land use change on ecosystem services in China. *Environ Monit Assess* 2016;**188**(4):1–13.
38. Buyantuyev A, Wu J. Urban heat islands and landscape heterogeneity: linking spatiotemporal variations in surface temperatures to land-cover and socioeconomic patterns. *Landsc Ecol* 2010;**25**:17–33.
39. Munafò M. *Consumo di suolo, dinamiche territoriali e servizi ecosistemici, 2021*. Report SNPA 22/21. Rome, Italy: ISPRA; 2021.
40. Romano B, Zullo F, Fiorini L, Ciabò S, Marucci A. Sprinkling: an approach to describe urbanization dynamics in Italy. *Sustainability* 2017;**9**(1):97.
41. Romano B, Zullo F, Fiorini L, Marucci A, Ciabò S. Land transformation of Italy due to half a century of urbanization. *Land Use Policy* 2017;**67**:387–400.
42. Ciommi M, Chelli FM, Carlucci M, Salvati L. Urban growth and demographic dynamics in southern Europe: oward a new statistical approach to regional science. *Sustainability* 2018;**10**(8):2765.
43. European Environmental Agency—EEA. *Urban sprawl in Europe—The ignored challenge. EEA Report no. 10*. Copenhagen, Denmark: European Environmental Agency; 2006.
44. European Environmental Agency—EEA. *Urban sprawl in Europe. EEA Report no. 11*. Copenhagen, Denmark: European Environmental Agency; 2016.
45. Podmanicky L, Balazs K, Belenyesi M, Centeri CS, Kristo'f D, Kohlheb N. Modelling soil quality changes in Europe. An impact assessment of land use change on soil quality in Europe. *Ecol Indic* 2011;**11**:4–15.
46. Salvati L, Zitti M. Territorial disparities, natural resource distribution, and land degradation: a case study in southern Europe. *Geojournal* 2007;**70**(2–3):185–94.
47. Leontidou L. *The Mediterranean City in transition: social change and urban development*. Cambridge, UK: Cambridge University Press; 1990.
48. Assennato F, Congedo L, Luti T, Marinosci I, Riitano N, Munafò M. Forme di urbanizzazione e dispersione urbana. In: Rapporto ISPRA 248, editor. *Consumo di suolo, dinamiche territoriali e servizi ecosistemici*. ISPRA: Rome, Italy; 2016.

49. Catalàn B, Sauri D, Serra P. Urban sprawl in the Mediterranean? Patterns of growth and change in the Barcelona metropolitan region 1993-2000. *Landsc Urban Plan* 2008;**85** (3–4):174–84.
50. Chorianopoulos I, Pagonis T, Koukoulas S, Drymoniti S. Planning, competitiveness and sprawl in the Mediterranean city: the case of Athens. *Cities* 2010;**27**:249–59.
51. Couch C, Petschel-held G, Leontidou L. *Urban sprawl in Europe: landscapes, land-use change and policy.* London, UK: Blackwell; 2007.
52. Munafò M, Norero C, Sabbi A, Salvati L. Urban soil consumption in the growing city: a survey in Rome. *Scott Geogr J* 2010;**126**(3):153–61.
53. Muñoz F. Lock living: urban sprawl in Mediterranean cities. *Cities* 2003;**20**(6):381–5.
54. Nickayin SS, Perrone F, Ermini B, Quaranta G, Salvia R, Gambella F, et al. Soil quality and peri-urban expansion of cities: a Mediterranean experience (Athens, Greece). *Sustainability* 2021;**13**(4):2042.
55. Romano B, Fiorini L, Marucci A, Zullo F. The urbanization run-up in Italy: from a qualitative goal in the boom decades to the present and future unsustainability. *Landarzt* 2020;**9**(9):301.
56. Weber CA, Puissant A. Urbanization pressure and modeling of urban growth: example of the Tunis metropolitan area. *Remote Sens Environ* 2003;**86**:341–52.
57. Zambon I, Benedetti A, Ferrara C, Salvati L. Soil matters? A multivariate analysis of socioeconomic constraints to urban expansion in Mediterranean Europe. *Ecol Econ* 2018;**146**:173–83.
58. Zambon I, Serra P, Sauri D, Carlucci M, Salvati L. Beyond the 'Mediterranean city': socioeconomic disparities and urban sprawl in three southern European cities. *Geogr Ann B: Hum Geogr* 2017;**99**(3):319–37.
59. Imura H, Chen J, Kaneko S, Matsumoto T. Analysis of industrialization, urbanization and land-use change in East Asia according to the DPSER framework. In: Center for Global Environmental Research, National Institute for Environmental Studies (Onogawa, Japan), editor. *Proceedings of NIES workshop on information bases and modeling for land-use and land-cover changes in East Asia, 25–27 January 1999*; 1999. p. 227–34.
60. McBratney A, Field DJ, Koch A. The dimensions of soil security. *Geoderma* 2014;**213**:203–13.
61. Cerreta M, De Toro P. Urbanization suitability maps: a dynamic spatial decision support system for sustainable land use. *Earth Syst Dynam* 2012;**3**:157–71.
62. Salvati L, Gemmiti R, Perini L. Land degradation in Mediterranean urban areas: an unexplored link with planning? *Area* 2012;**44**(3):317–25.
63. Tuttitalia. Città Metropolitana di Roma Capitale; 2021. [10/11/2021], Available at https://www.tuttitalia.it/lazio/provincia-di-roma/.
64. Salvati L, Sabbi A. Exploring long-term land cover changes in an urban region of southern Europe. *Int J Sust Dev World* 2011;**18**(4):273–82.
65. Frondoni R, Mollo B, Capotorti G. A landscape analysis of land cover change in the municipality of Rome (Italy): Spatio-temporal characteristics and ecological implications of land cover transitions from 1954 to 2001. *Landsc Urban Plan* 2011;**100** (1–2):117–28.
66. European Environmental Agency—EEA. *Mapping guide for a European urban atlas. Version 1.1.* Copenhagen, Denmark: European Environmental Agency; 2010.
67. *Urban Atlas (Copernicus Land Monitoring Service—CLMS), [12/09/2022]*; 2018. Available at https://land.copernicus.eu/local/urban-atlas.
68. European Commission—EC. *Towards a thematic strategy for soil protection. COM(2002) 179 final.* Bruxelles, Belgium: European Commission; 2002.
69. European Commission—EC. *EU Soil Strategy for 2030. Reaping the benefits of healthy soils for people, food, nature and climate. COM(2021) 699 final.* Bruxelles, Belgium: European Commission; 2021.

70. Italian Institute for Environmental Protection and Research—ISPRA; 2008. [15/09/2022], Available at: https://www.isprambiente.gov.it/en/publications/technical-periodicals/descriptive-memories-of-the-geological-map-of/the-geology-of-rome.-from-the-historical-center-to-the-outskirts.
71. Arnoldus-Huyzendveld A. *I Suoli di Roma*. Rome, Italy: Municipality of Rome, Dip.to X—II° U.O. Sviluppo Sostenibile; 2003.
72. Biasi R, Colantoni A, Ferrara C, Ranalli F, Salvati L. In-between sprawl and fires: long-term forest expansion and settlement dynamics at the wildland–urban interface in Rome, Italy. *Int J Sust Dev World* 2015;**22**(6):467–75.
73. Ferrara C, Salvati L, Tombolini I. An integrated evaluation of soil resource depletion from diachronic settlement maps and soil cartography in peri-urban Rome, Italy. *Geoderma* 2014;**232**:394–405.
74. Salvati L, Zitti M, Sateriano A. Changes in city vertical profile as an indicator of sprawl: eEvidence from a Mediterranean urban region. *Habitat Int* 2013;**38**:119–25.
75. Munafò M. *Consumo di suolo, dinamiche territoriali e servizi ecosistemici*. Report SNPA 32/22. Rome, Italy: ISPRA; 2022.
76. Riitano N, Congedo L, Garofalo V, La Mantia C, Luti T, Marinosci I, et al. Stima del consumo di suolo a livello nazionale e regionale. In: ISPRA, editor. *Consumo di suolo, dinamiche territoriali e servizi ecosistemici. Rapporto ISPRA 248*. Rome, Italy: ISPRA; 2016. p. 12–6.
77. Bren d'Amour CB, Reitsma F, Baiocchi G, Barthel S, Güneralp B, Erb KH, et al. Future urban land expansion and implications for global croplands. *Proc Natl Acad Sci U S A* 2017;**114**(34):8939–44.
78. Ceccarelli T, Bajocco S, Perini L, Salvati L. Urbanisation and land take of high quality agricultural soils—exploring long-term land use changes and land capability in northern Italy. *Int J Environ Res* 2014;**8**(1):181–92.
79. Salomon MJ, Watts-Williams SJ, McLaughlin MJ, Cavagnaro TR. Urban soil health: a city-wide survey of chemical and biological properties of urban agriculture soils. *J Clean Prod* 2020;**275**, 122900.
80. Salvati L, Ciommi MT, Serra P, Chelli FM. Exploring the spatial structure of housing prices under economic expansion and stagnation: the role of socio-demographic factors in metropolitan Rome, Italy. *Land Use Policy* 2019;**81**:143–52.
81. Salvati L. Towards a polycentric region? The socio-economic trajectory of Rome, an 'eternally Mediterranean' city. *Tijdschr Econ Soc Geogr* 2014;**105**(3):268–84.
82. Perrone F. *Per il governo del consumo di suolo. Strumenti della pianificazione territoriale e paesaggistica per il controllo dei processi di artificializzazione dei suoli*. Rome, Italy: Sapienza University of Rome—Dip. PDTA; 2018.
83. Aguilar AG. Peri-urbanization, illegal settlements and environmental impact in Mexico City. *Cities* 2008;**25**:133–45.
84. Zavala MA, Burkey TV. Application of ecological models to landscape planning: the case of the Mediterranean basin. *Landsc Urban Plan* 1997;**38**:213–27.
85. Jabareen YR. Sustainable urban forms: their typologies, models, and concepts. *J Plan Educ Res* 2006;**26**(1):38–52.
86. Garcia D, Riera P. Expansion versus density in Barcelona: a valuation exercise. *Urban Stud* 2003;**40**(10):1925–36.
87. Marull J, Pino J, Tello E, Cordobilla MJ. Social metabolism, landscape change and land-use planning in the Barcelona metropolitan region. *Land-use Pol* 2009;**27**(2):497–510.
88. Paul V, Tonts M. Containing urban sprawl: trends in land-use and spatial planning in the metropolitan region of Barcelona. *J Environ Plan Manag* 2005;**48**(1):7–35.
89. Piano Regolatore Generale—PRG. *Relazione. Rome, Italy: Municipality of Rome*; 2008.

90. Piano Regolatore Generale—PRG. *Norme Tecniche di Attuazione. Rome, Italy: Municipality of Rome*; 2008.
91. Larson WE, Pierce FJ. The dynamics of soil quality as a measure of sustainable management. In: Doran J, Coleman D, Bezdicek D, Stewart B, editors. *Defining soil quality for a sustainable environment.* Fitchburg, USA: SSSA Special Publications; 1994. p. 37–51.

CHAPTER SIX

Mapping and assessment of flood regulation supply and demand in Vilnius (Lithuania)

Paulo Pereira[a,*], Marius Kalinauskas[a], Manob Das[b], Katažyna Bogdzevič[a], Miguel Inácio[a], and Damia Barcelo[c]
[a]Environmental Management Laboratory, Mykolas Romeris University, Vilnius, Lithuania
[b]Department of Geography, University of Gour Banga, Malda, West Bengal, India
[c]Catalan Institute for Water Research (ICRA-CERCA), Girona, Catalonia, Spain
[*]Corresponding author: e-mail address: pereiraub@gmail.com

Contents

Abstract

Heavy rain events are expected to be more frequent and intense due to climate change, which will likely increase the vulnerability of the urban area to flood episodes. In this context, it is crucial to identify the areas with a high capacity to retain water and the areas where flood protection is more needed. To address this problem, this work aims to map and assess the flood regulation supply and demand ecosystem services in Vilnius, Lithuania. The results showed that Vilnius has a large area with a flood regulation supply, especially in the northeast and southern areas. The areas with high flood regulation demand were located in the old town and several urbanizations scattered inside the city limits. We used the soil-vegetation adjusted index (SAVI) to validate the flood regulation supply index using linear regression. In contrast, the flood regulation demand index was validated by comparing the historical flood points with the model results.

Advances in Chemical Pollution, Environmental Management and Protection, Volume 8

ISSN 2468-9289
https://doi.org/10.1016/bs.apmp.2022.10.015

SAVI predicted acceptably ($r2 = 0.43$), and the flooded points identified were located in the areas with high flood regulation demand index. Both models were mismatched (−0.50. $P < 0.001$), showing that they have different spatial patterns. These results obtained in this work can help planners and decision-makers better plan the areas with a high capacity to retain water and identify the areas where the demand is high.

Keywords: Floods, Ecosystem services, Climate change, Supply, Demand

1. Introduction

Floods are one of the most destructive events. According to WHO,[1] between 1970 and 2019, floods were responsible for 58,700 deaths and an economic loss of US$ 115 billion. The existent climate forecasts that these events will be more frequent and severe in the future due to climate change (e.g., glacier melting, heavy rainfall, hurricanes).[2–4] For instance, the IPCC[5] highlighted that heavy rainfall would increase with high confidence in large parts of the globe. A warmer climate will increase the very wet events and the intensification of storm tracks, cyclones, and hurricanes. On the other hand, a warmer climate will increase glacier melting and the flooding events associated with these events.[5] Although climate change is increasing the intensity and frequency of flood weather-related events, land use change has a vital role in increasing the exposure and vulnerability of the population to floods.[6,7] Urban development (e.g., soil sealing) is one factor that dramatically increases the impact of floods. The impermeable surfaces reduce water infiltration and increase the overland flow. This enhances flood destruction, especially in areas with heterogeneous topography.[8,9] Urban areas, especially those near the coast and rivers, are vulnerable to flood events.[10,11]

Habitat change (e.g., urbanization) is one of the main drivers of change in ecosystem services loss. Soil sealing hampers the capacity of the ecosystems to supply regulating (e.g., air quality, flood, erosion, climate and water purification), provisioning (e.g., food, fiber) and cultural (e.g., landscape esthetics), and ecosystem services.[12] Among these services, flood regulation is one of the most relevant since it can reduce costs and live losses. In urban areas, there is a high demand for flood regulation, especially in areas with a high population density.[13,14] Traditionally, flood regulation has been carried out using hard engineering approaches (e.g., sewage systems). Recently, several researchers have called attention to the importance of nature-based solutions (e.g., urban parks, gardens, wetlands) and their relevance for flood regulation and reducing their impact on human assets.[15–17]

Urban green areas are essential to increase human well-being and make urban areas more liveable.[18] Therefore, it is vital to identify areas of flood regulation supply and demand in urban areas to know where water can be stored with more efficiency and where the vulnerability is high.[19,20] Mapping is an excellent exercise to spatialize flood supply and demand since it efficiently synthesizes the information for planners and decision-makers to better manage it.[21] Several works were carried out about flood regulation supply[22,23] and demand[24,25] in urban areas. Nevertheless, most of them were done using simple and qualitative methods (e.g., expert ranking matrix).[26,27] Recently, studies started to become more quantitative, and numerous models (e.g., InVEST) have been developed to map flood regulation.[28,29] Although this represents a significant advance, most quantitative models lack validation, which is key for the credibility of the outputs and assessing the uncertainty and limitations.[30] Considering the importance of floods in urban environments and the relevance of mapping and assessing flood regulation supply and demand with accuracy, this chapter aims to develop a quantitative and validated framework to map and assess flood supply and demand in Vilnius, Lithuania.

2. Materials and methods

2.1 Study area

Vilnius is Lithuania's capital, located at 54.6872° N, 25.2797° E (Fig. 1). The average annual temperature is 8.1 °C, and precipitation is 735 mm (2016–2020).[31,32] In 2021, the city had 588,412 inhabitants. The population density is 1337 inhabitants/km^2. In 1984, the city center was classified as part of the world heritage list.[33] According to the Husqvarna Urban Green Space Index,[34] Vilnius is one of the greenest cities in Europe. The green areas occupy 18.4% of the municipal territory and have many recreation areas.[35,36]

2.2 Data and flood regulation supply and demand framework

The data used in this work is shown in Table 1. To map and assess flood regulation, we create an index based on different variables. All the data was resampled to 20 m resolution using ArcMap 10.2 resampling tool. We used soil texture to estimate the capacity of soil infiltration,[37] tree density, grassland cover, and imperviousness to identify the areas with a high capacity of water retention.[38] Soil thickness and depth to the impermeable layer assess the soil's capacity to store water.[39] Regarding flood regulation

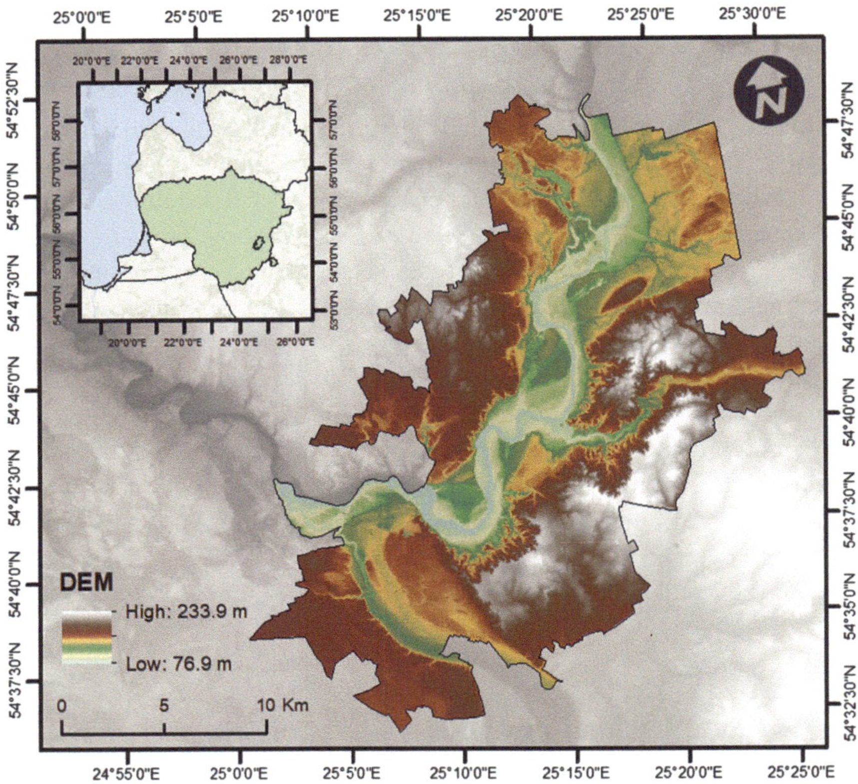

Fig. 1 Study area.

demand, we assessed the slope (20 m resolution) to identify the areas where floods have a high destruction capacity.[40] The slope was calculated using ArcMap 10.2 spatial analyst tool. The topographic Wetness index (TWI) was used to evaluate the areas more vulnerable to water accumulation.[41] TWI was calculated using QGIS 2.6 TWI plugin. Water line density, distance to lakes and water lines were used to identify the areas more prone to flooding.[42] Water line density was calculated using ArcMap 10.2 line density tool, and distance to lakes and distance to water lines using ArcMap 10.2 Euclidean distance tool. Flooding area (return period 1000 years) was used to identify the areas more prone to flooding and imperviousness to identify the areas with higher soil sealing.[43] Finally, building and population density were assessed to identify the areas with high human infrastructure and population, respectively. These variables were calculated using ArcMap 10.2 Kernel density tool.

Table 1 Data used.

Variables	Units	Period	Data type	Resolution (m)	Reference
Digital elevation model	–	–	Raster	10	https://aaa.lrv.lt/
Tree density	%	2018		20	https://land.copernicus.eu
Grassland					
Imperviousness					
Population density	km^2	2015		20	https://www.eea.europa.eu/
Sentinel 2 images		2015i–2020		20	https://www.esa.int
Soil thickness	cm		Vector	–	https://esdac.jrc.ec.europa.eu/
Soil depth impermeable layer	cm			–	
Soil texture	%			–	https://esdac.jrc.ec.europa.eu/
Water bodies	–	–		–	https://www.registrucentras.lt
Buildings	–	–		–	
Flood return period (1000 years)	–	–		–	https://aaa.lrv.lt/
Historical floods	–	1931–2020		–	https://archivesofculture.com/ https://www.15min.lt/ https://www.delfi.lt/

Multicollinearity was assessed for the variables of the flood retention supply and demand model (Tables 2 and 3). Multicollinearity was considered at an $r > 0.7$.[44] In the current case, no multicollinearity was identified. Therefore, all the variables were considered in the model. In both models, the variables were reclassified between 0 and 1 and overlaid. In the flood regulation supply model, we applied different weights to the variables (Fig. 2). We applied a higher weight to imperviousness (0.4) because the capacity of an area to retain floods depends on the degree of soil sealing.

Table 2 Multicolinearity results for the variables used in the flood regulation supply index.

Variable	Slope	Soil thickness	Soil DIL	Soil texture	Tree_density	Grassland	Impervisioness
Slope	–						
Soil thickness	0.003	–					
Soil DIL	−0.001	−0.342	–				
Soil texture	−0.120	−0.151	0.604	–			
Tree_density	0.178	−0.098	0.017	−0.045	–		
Grassland	−0.074	0.067	−0.047	−0.005	−0.314	–	
Impervisioness	−0.145	0.050	0.042	0.062	−0.424	−0.081	–

Table 3 Multicolinearity results for the variables used in the flood regulation demand index.

Variable	Slope	TWI	Dist. to rivers	River density	Dist. to lakes	Impervisioness	Flood (1000)	Buil. density	Pop. density
Slope	–								
TWI	0.020	–							
Dist. to rivers	0.032	0.059	–						
River density	−0.006	−0.108	−0.151	–					
Dist. to lakes	0.071	−0.024	0.019	0.057	–				
Impervisioness	−0.145	−0.011	−0.046	0.115	0.167	–			
Flood (1000)	−0.026	−0.067	−0.167	0.125	0.010	−0.079	–		
Buil. density	0.099	0.027	−0.026	0.182	0.065	0.384	0.025	–	
Pop. density	0.033	0.014	−0.006	0.105	0.191	0.529	−0.130	0.480	–

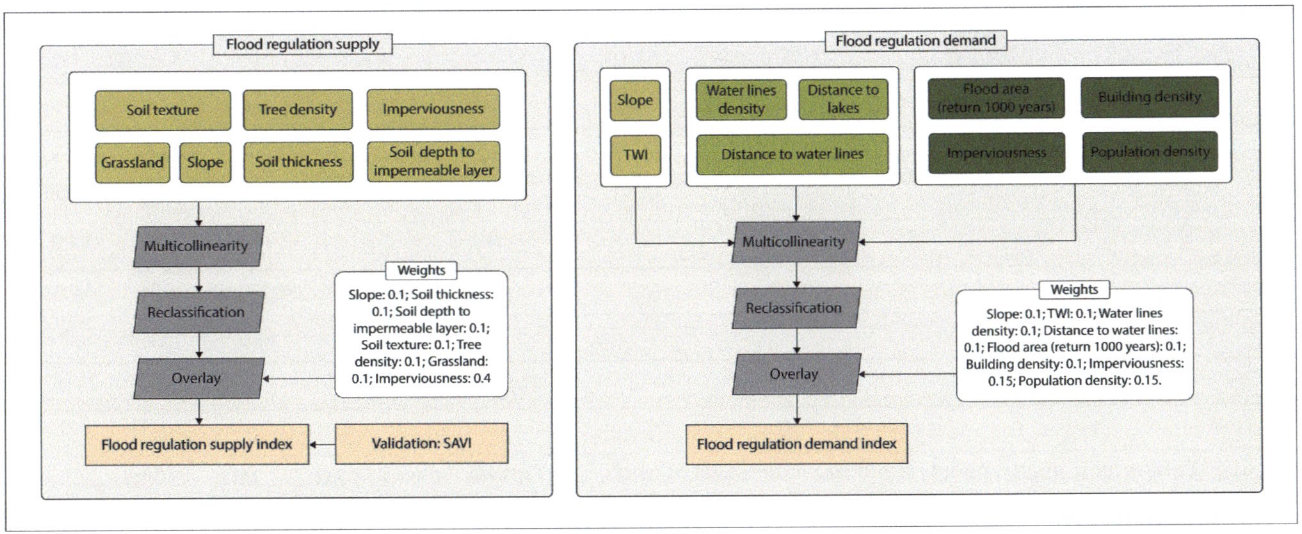

Fig. 2 Framework to assess flood regulation supply and demand.

If the soil is sealed, irrespective of the other characteristics (e.g., slope, texture, thickness), the capacity to retain water is hampered. To the other variables, we applied weight of 0.1 (Fig. 2). In urban areas, soil sealing is vital for flood development.[43] Regarding the flood regulation model, it is important to identify where the soil has a high degree of sealing (e.g., imperviousness) and the population is located because the demand can be high in these areas. In this context, we applied a higher weight to imperviousness and population density (0.15) than to the other variables (0.1). For instance, it is well known that the areas with high population density are the ones with the highest demand for flood protection.[45] For mapping, data was reclassified in 10 classes using the natural breaks (Jenks) method. The flood regulation model was validated using the soil-vegetation adjusted index (SAVI), and assessed using Sentinel 2 images (Table 1). SAVI describes the dynamic between soil and vegetation.[46] This index ranges between −1 and 1. A low SAVI index shows that the soil and the vegetation are in poor conditions, while a high value shows the opposite.[47] SAVI was calculated using the following formula:

$$SAVI = \left[\frac{NIR - RED}{NIR + RED + 0.5}\right] * (1 + 0.5) \quad (1)$$

Where, NIR is the Near Infra-Red band 8 and RED is the band 5. Cloud-free images from March–August (2015–2021) were used. The data for each month were averaged using ArcMap 10.2 raster calculator tool. After averaging all the monthly images for the period between 2015 and 2021, we calculated the mean value for the period between March and August. To validate the flood regulation demand index, we consulted historical archives (1931–2020) from where floods were identified. The data was vectorized using ArcMap 10.2.

2.3 Statistical analysis

A grid with 1580 randomly distributed points was created to extract data for statistical analysis. The average values of the flood regulation supply and demand and SAVI were carried out using ArcMap 10.2 extract values to points tool. The same procedure was applied to extract the values of the points where floods were identified (Flood Regulation demand index). Model validation and the relation between flood regulation supply and demand indexes were assessed with linear regression and a Spearman correlation coefficient, and significant correlations were considered at a $P < 0.05$. Statistical analyses were carried out using JASP 0.16.3 and graphics with online Plotly (https://chart-studio.plotly.com/).

3. Results and discussion

3.1 Flood regulation supply and demand

The results showed that, on average, the flood regulation supply index was 0.490, and the flood regulation demand index was 0.475. The standard deviation of both models was low: 0.085 for the flood regulation supply index and 0.069 for the flood regulation demand index (Table 4). In the flood regulation supply index, the number of pixels was observed in the class between 0.57 and 0.62 (Fig. 3A), while in the flood regulation demand index, the highest number of pixels was observed in the class between 0.4 and 0.42 (Fig. 3B). The areas where the flood regulation supply index had the highest values were located in the northeast and south parts of Vilnius (Fig. 3A). Contrarily, the flood regulation demand index had the highest values in the central area (old town) and several districts scattered in the north, east and southeast areas of Vilnius (Fig. 3B). The areas with the highest values of flood regulation supply were associated with the presence of forests and urban parks,[35] while the lowest are located in areas with high imperviousness. The results were similar to those observed in other urban areas, such as Shenyang-China,[48] Guangzhou-China,[49] Beijing-China,[50] Bandung City-Indonesia[24] and Dolo-Italy[51] where urban green areas had a high capacity for flood retention. The flood regulation demand was high in the old town, and some scattered in different parts of the city, where an urbanization process has been observed in recent years. Previous works observed that old towns have a high vulnerability to floods due to dense urbanization[52] as in the case of Venice-Italy,[53] Badalona-Spain[54] and Seoul-South Korea.[55] As identified in this work, the expansion of urban areas increases the demand for protection

Table 4 Descriptive statistics.

	Flood regulation supply index	**Flood regulation demand index**	**Flooding points (flood regulation demand index)**
Median	0.510	0.465	0.540
Mean	0.490	0.475	0.543
Std. deviation	0.085	0.069	0.053
Minimum	0.200	0.305	0.420
Maximum	0.590	0.700	0.650

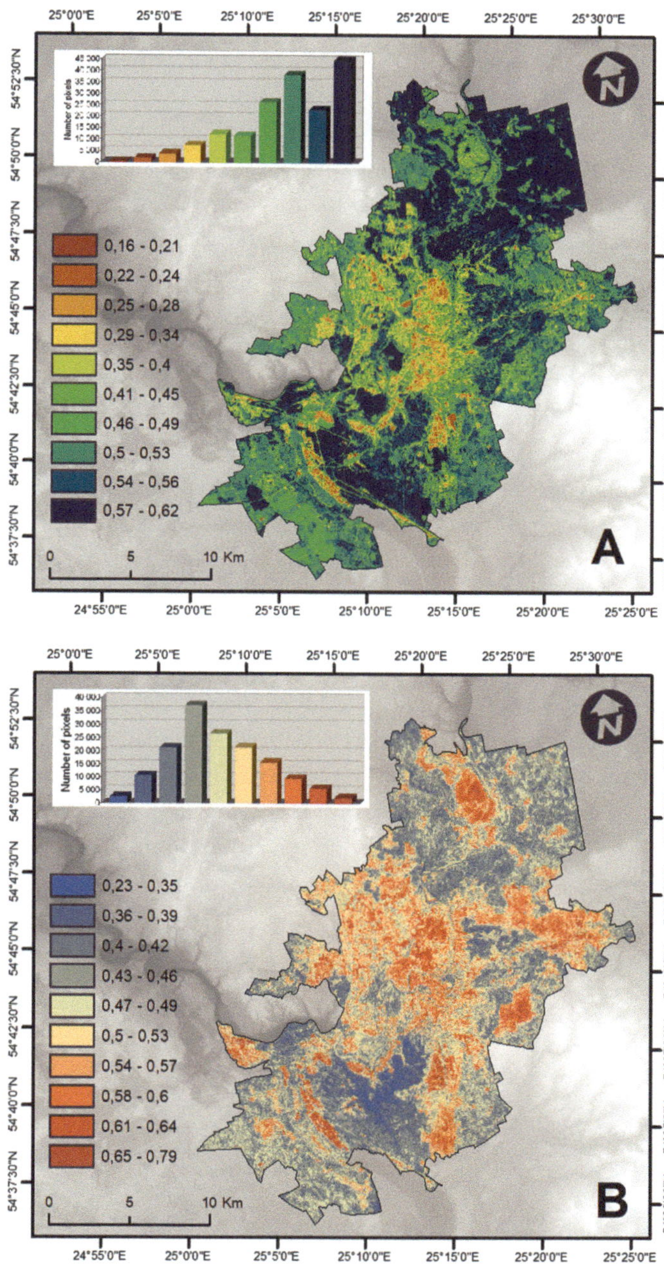

Fig. 3 (A) Flood supply index and (B) flood demand index.

due to population expansion and impervious surfaces.[56,57] The flood regulation supply index validation is shown in Fig. 4. The coefficient of determination between the index and SAVI was 0.43, and the correlation of 0.58, $P < 0.001$. Flood regulation supply depends on multiple factors (e.g., soil properties and topography). SAVI explained 43% of the flood regulation supply. Similar coefficients of determination were found in the validation of other ecosystem services such as coastal recreation,[58] habitat quality[59] and seafood.[60] Regarding the flood regulation demand model, comparing the results with the flooded points retrieved from the archives (Fig. 5), it is visible that the dots match the areas with high flood regulation demand. On average, the values of the flooding points extracted from the flood regulation demand model had an average of 0.543, a minimum of 0.420 and a maximum of 0.650. Most of the values observed were higher than 0.54, showing that they were identified in areas where the flood regulation demand is high (Fig. 6). Overall, the areas where this index is high match the points where the historical floods were identified. The mismatch between the flood regulation supply and demand indexes is shown in Fig. 7. The relation between them is relative (-0.58, $P < 0.001$). The mismatch between the indexes was expected since the areas with a high capacity for flood retention are covered by vegetation and have a reduced imperviousness and population density. The contrary occurs in the areas with high demand, which generally have high urbanization and population density. This mismatch between ecosystem services supply and demand components was also observed in previous works.[61,62]

3.2 Uncertainty and limitations

Different data resolutions in all models developed are a concern.[59] In ours, it was an important issue as well. We tried to access data with the highest

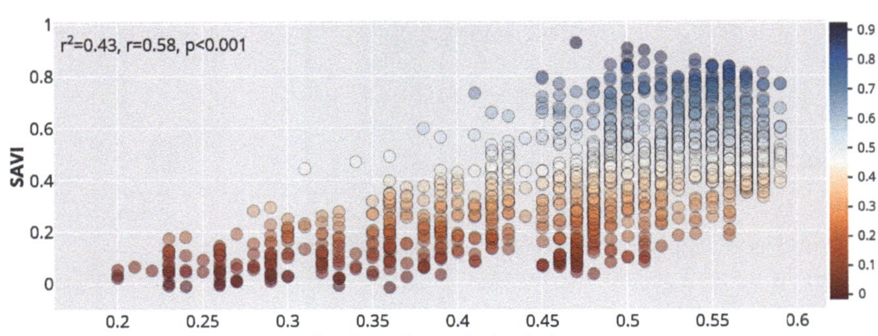

Fig. 4 Relation between flood supply regulation index and SAVI.

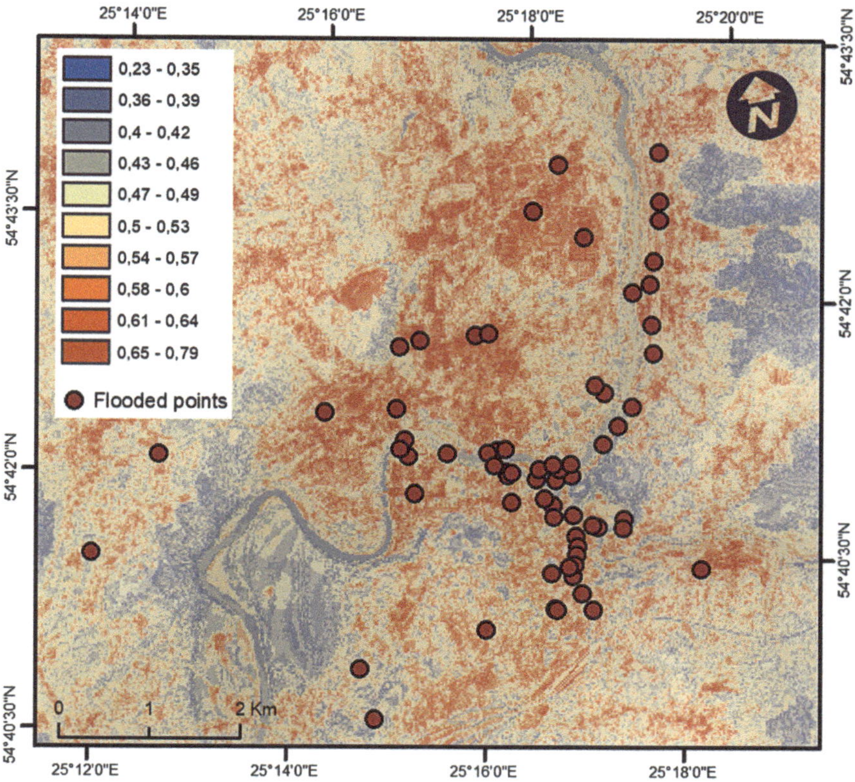

Fig. 5 Observed flooded points and flood demand index.

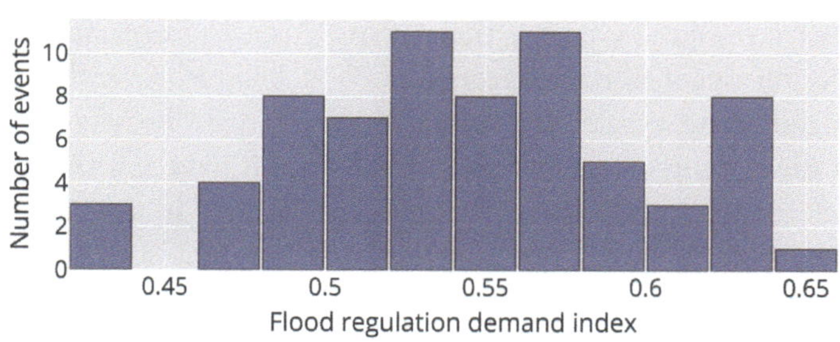

Fig. 6 Histogram of the values extracted (flood regulation demand index) from the flooded points.

resolution possible. Except for the digital elevation model (10 m) (Table 1), all data had a resolution of 20 m. To overlay the models correctly, we had to resample the digital elevation model data to 20 m, and we may have loosed some information. Nevertheless, for the size of Vilnius (400.54 km^2), having a model of 20 m is accurate. We used a remote sensing index (SAVI) to

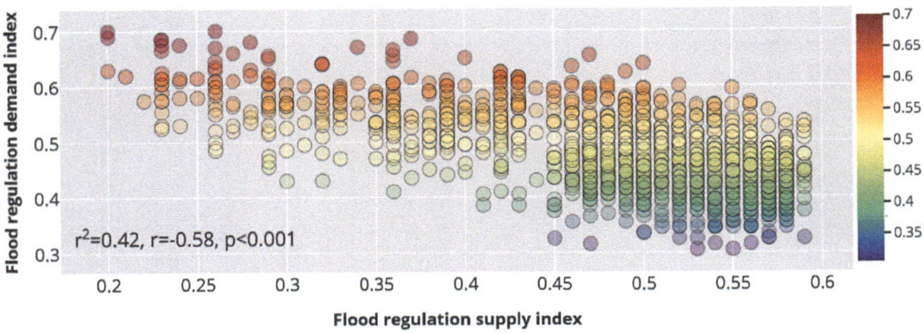

Fig. 7 Relation between flood regulation supply index and flood regulation demand index.

model the flood regulation supply, which may have some inaccuracies. We averaged the SAVI value to understand the soil-vegetation dynamic globally. In the case of a flood event, the soil-vegetation properties cannot be the same that we assumed here. For instance, the antecedent soil water content[63,64] and hydrophobicity[65] before the heavy rainfall event have an important impact on flood development. In this model, we did not consider these variables. In addition, they have high spatio-temporal variability,[66,67] increasing the difficulty of measuring them accurately. The Sentinel 2 images are only available from 2015, and this short analysis time may also pose some limitations in SAVI estimation. A larger series of images would be key to have. We decided to use Sentinel 2 images because the resolution was similar to the other indicators, and the modeling errors would be minimal. Other satellite images such as LANDSAT could be used. However, the majority of the bands have a resolution higher than 30 m.[a] Also, the capacity of the soil to store water depends on management.[68] In Vilnius, several urban areas are managed using machinery, and the soils can be heavily compacted, reducing their ability to store water.[69] We do not have information about the management applied in urban gardens, which is also a limitation. Overall in future studies, ground-truth (e.g., soil and management approach) data need to be used to increase the accuracy of the flood regulation supply index. Although the model has some limitations, the validation was acceptable ($R^2 = 0.43$). Regarding the flood regulation demanding model, the identification of flooded points was limited to the sources available (e.g., historical archives). The information was very scattered, and likely more areas could be affected by floods that we could not identify. We found data back to 1931. Early information was not available. In any case, the

[a] https://www.usgs.gov/faqs/what-are-band-designations-landsat-satellites

flooded areas matched the areas where the flood regulation demand index was the highest. Urban areas' land use change is very dynamic.[70] The models we performed were static in time, which is a limitation of the effect of the urbanization process on flood regulation supply and demand indexes. Further works need to consider a temporal approach. However, this depends on the data availability.

4. Implications for management

Vilnius is one of the greenest cities in the world, although several urban areas are being affected by a rapid sprawl.[35,71] The flood regulation index results showed that the city has a high capacity to regulate floods, which is important to maintain. This is key to increasing resilience to the projected increase of heavy rainfall periods in a climate change context.[5] Limiting the sprawl of new urbanizations is necessary since they can increase the vulnerability to floods. Local authorities need to promote the development of a compact city and limit urban expansion in areas with a high capacity to retain water. Nevertheless, the development of compact cities may pose some challenges regarding flood risk because they can be more vulnerable due to the concentration of population and the impervious surface.[72,73] In this case, it is key to develop good water storage systems or establish nature-based solutions that can retain water (e.g., urban green areas). Several spaces of Vilnius are natural areas (e.g., Verkiu and Pavilnys regional natural parks), and urban expansion in these areas is forbidden. However, the pressure on the surrounding areas is becoming high,[74] which can potentially affect these areas' capacity to retain water. Vilnius's old town has a very high flood regulation demand, and most of the flood events were identified in this area. In order to reduce the impacts, several solutions have been applied (e.g., increasing runoff storage capacity). Nevertheless, these efforts did not decrease spring and summer floods. The city still has water drainage problems. Therefore, although it might be challenging to apply measures based on a nature-based solution approach (e.g., permeable areas such as permeable pavements or urban green areas) in the city center due to the land price and the lack of space, the consideration of more efficient measures to retain water could be considered. The expansion of new urbanizations in Vilnius is increasing the flood regulation demand and reducing the supply. These developments should be accompanied by infrastructures that could reduce the risk of floods based on nature-based solutions, such as the ones mentioned above.

5. Conclusions

Climate change is increasing the severity and frequency of heavy rain episodes. This will be important in the flood development in urban areas due to the high sealed area. Currently, Vilnius has many areas with a high flood regulation supply model, mainly in the northeast and south parts of the city. The areas with the highest flood regulation demand model were identified in the old town and several neighborhoods scattered inside the urban area perimeter, some of them comprising new developments. If the urban expansion continues, flood regulation demand will likely increase. To assess flood supply and demand, a model framework was developed. Overall, the flood regulation supply and demand model's validation were acceptable. SAVI predicted a good flood regulation supply index, and the historical flood points matched the areas with the highest flood regulation demand index. As expected, both models had a mismatch, showing that the areas with the highest flood regulation supply (e.g., forests) are different from the ones where the flood regulation demand index had the highest values (e.g., high soil sealing and population density).

Acknowledgments

This work was developed within the Lithuanian National Ecosystem Services Assessment and Mapping (LINESAM), which has received funding from European Social Fund project LINESAM no. 09.3.3-LMT-K-712-01-0104 under a grant agreement with the Research Council of Lithuania (LMTLT).

References

1. WHO. *WHO atlas of mortality and economic losses from weather, climate and weather extremes (1970–2019)*. Geneva: World Meteorological Organization (WMO); 2019.
2. Hosseinzadehtalaei P, Tabari H, Willems P. Climate change impact on short-duration extreme precipitation and intensity–duration–frequency curves over Europe. *J Hydrol* 2020;**590**, 125249. https://doi.org/10.1016/j.jhydrol.2020.125249.
3. Marsooli R, Lin N, Emanuel K, Feng K. Climate change exacerbates hurricane flood hazards along US Atlantic and Gulf Coasts in spatially varying patterns. *Nat Commun* 2019;**10**:3785. https://doi.org/10.1038/s41467-019-11755-z.
4. Zheng G, Allen SK, Bao A, Ballesteros-Cánovas JA, Huss M, Zhang G, et al. Increasing risk of glacial lake outburst floods from future third pole deglaciation. *Nat Clim Chang* 2021;**11**:411–7. https://doi.org/10.1038/s41558-021-01028-3.
5. IPCC. Summary for policymakers. In: Masson-Delmotte V, Zhai P, Pirani A, Connors SL, Péan C, Berger S, Zhou B, editors. *Climate change 2021: the physical science basis. contribution of working group I to the sixth assessment report of the intergovernmental panel on climate change*; 2021.

6. Gao Y, Chen J, Luo H, Wang H. Prediction of hydrological responses to land use change. *Sci Total Environ* 2020;**708**:134998. https://doi.org/10.1016/j.scitotenv.2019.134998.
7. Pereira P, Inacio M, Kalinauskas M, Bogdzevič K, Bogunovic I, Zhao W. Land-use changes and ecosystem services. In: Pereira P, Gomes E, Rocha J, editors. *Mapping and forecast land use/cover changes.* The Present and Future of Planning; 2022. p. 1–27. https://doi.org/10.1016/B978-0-323-90947-1.00007-7.
8. Gabriels K, Willems P, Van Orshoven J. An iterative runoff propagation approach to identify priority locations for land cover change minimizing downstream river flood hazard. *Landsc Urban Plan* 2022;**218**:104262. https://doi.org/10.1016/j.landurbplan.2021.104262.
9. Recanatesi F, Petroselli A. Land cover change and flood risk in a peri-urban environment of the metropolitan area of Rome (Italy). *Water Resour Manag* 2020;**34**:4399–413. https://doi.org/10.1007/s11269-020-02567-8.
10. Abebe Y, Kabir G, Tesfamariam S. Assessing urban areas vulnerability to pluvial flooding using GIS applications and Bayesian Belief Network model. *J Clean Prod* 2018;**174**:1629–41. https://doi.org/10.1016/j.jclepro.2017.11.066.
11. Cho SY, Chang H. Recent research approach. es to urban flood vulnerability, 2006–2016. *Nat Hazards* 2017;**88**:633–49. https://doi.org/10.1007/s11069-017-2869-4.
12. Pereira P. Ecosystem services in a changing environment. *Sci Total Environ* 2020;**702**:135008. https://doi.org/10.1016/j.scitotenv.2019.135008.
13. O'Donnell EC, Thorne CR. Drivers of future urban flood risk. *Philos Trans R Soc* 2020;**378**:20190216. https://doi.org/10.1098/rsta.2019.0216.
14. Pinto LV, Pereira P, Gazdic M, Ferreira A, Ferreira CSS. Assessment of NBS impact on pluvial flood regulation within urban areas: a case study in Coimbra, Portugal. In: Ferreira CSS, Kalantari Z, Hartmann T, Pereira P, editors. *Nature-based solutions for flood mitigation. The handbook of environmental chemistry.* vol 107. Cham: Springer; 2022. p. 289–312. https://doi.org/10.1007/698_2021_769.
15. Huang Y, Tian Z, Liu J, Irannezhad M, Fan D, Hou M, et al. Nature-based solutions for urban pluvial flood risk management. *WIRE's Water* 2020;**7**:e1421. https://doi.org/10.1002/wat2.1421.
16. Santoro S, Pluchinotta I, Pagano A, Pengal P, Cokan B, Giordano R. Assessing stakeholders' risk perception to promote nature based solutions as flood protection strategies: the case of the Glinščica river (Slovenia). *Sci Total Environ* 2020;**655**:188–201. https://doi.org/10.1016/j.scitotenv.2018.11.116.
17. Seddon N, Chausson A, Berry P, Girardin CAJ, Smith A, Turner B. Understanding the value and limits of nature-based solutions to climate change and other global challenges. *Philos Trans R Soc B* 2020;**375**:20190120. https://doi.org/10.1098/rstb.2019.0120.
18. Ferreira CSS, Kalantari Z, Pereira P. Liveable cities: current environmental challenges and paths for urban sustainbility. *J Environ Manag* 2021;**221**:111458.
19. Sing A, Sarma AK, Hack J. Cost-effective optimization of nature-based solutions for reducing urban floods considering limited space availability. *Environ Process* 2020;**7**:297–319. https://doi.org/10.1007/s40710-019-00420-8.
20. Zeleňáková M, Constantin Diacon UD, Haarstad K. Urban water retention measures. *Procedia Eng* 2017;**190**:419–26. https://doi.org/10.1016/j.proeng.2017.05.358.
21. Pereira P, Brevik E, Trevisani E. Editorial: Mapping the environment. *Sci Total Environ* 2018;**610–611**:17–23. https://doi.org/10.1016/j.scitotenv.2017.08.001.
22. Gunnell K, Mulligan M, Francis RA, Hole DG. Evaluating natural infrastructure for flood management within the watersheds of selected global cities. *Sci Total Environ* 2019;**670**:411–24. https://doi.org/10.1016/j.scitotenv.2019.03.212.
23. Li P, Sheng M, Yang D, Tang L. Evaluating flood regulation ecosystem services under climate, vegetation and reservoir influences. *Ecol Indic* 2019;**107**:105642. https://doi.org/10.1016/j.ecolind.2019.105642.

24. Afriyanie D, Julian MM, Riqqi A, Akbar R, Suroso DSA, Kustiwan I. Re-framing urban green spaces planning for flood protection through socio-ecological resilience in Bandung City, Indonesia. *Cities* 2020;**101**:102710. https://doi.org/10.1016/j.cities.2020.102710.
25. Shen J, Du S, Huang Q, Yin Q, Zhang M, Wen J, et al. Mapping the city-scale supply and demand of ecosystem flood regulation services—A case study in Shanghai. *Ecol Indic* 2019;**106**:105544. https://doi.org/10.1016/j.ecolind.2019.105544.
26. Tao Y, Wang H, Ou W, Guo J. A land-cover-based approach to assessing ecosystem services supply and demand dynamics in the rapidly urbanizing Yangtze River Delta region. *Land Use Policy* 2018;**72**:250–8. https://doi.org/10.1016/j.landusepol.2017.12.051.
27. Wang L, Zheng H, Wen Z, Liu L, Robinson BE, Li R, et al. Ecosystem service synergies/trade-offs informing the supply-demand match of ecosystem services: framework and application. *Ecosyst Serv* 2019;**37**:100939. https://doi.org/10.1016/j.ecoser.2019.100939.
28. Kadaverugu A, Rao CN, Viswanadh VK. Quantification of flood mitigation services by urban green spaces using InVEST model: a case study of Hyderabad city, India. *Model Earth Syst Environ* 2021;**7**:589–602. https://doi.org/10.1007/s40808-020-00937-0.
29. Pathak S, Liu M, Jato-Espino D, Zevenbergen C. Social, economic and environmental assessment of urban sub-catchment flood risks using a multi-criteria approach: A case study in Mumbai City, India. *J Hydrol* 2020;**591**:125216. https://doi.org/10.1016/j.jhydrol.2020.125216.
30. Inacio M, Karnauskaite D, Baltranaite E, Kalinauskas M, Bogdzevic K, Gomes E, et al. Ecosystem services of the Baltic Sea: an assessment and mapping perspective. *Geogr Sustain* 2020;**1**:256–65. https://doi.org/10.1016/j.geosus.2020.11.001.
31. Lithuanian Hydrometeorological Service. *Meteo.lt*; 2022. Retrieved February 14, 2022, from http://www.meteo.lt/krituliai. http://www.meteo.lt/krituliai.
32. Official Statistics Portal. *Average temperature*; 2020. Retrieved January 31, 2022, from https://osp.stat.gov.lt. https://osp.stat.gov.lt/statistiniu-rodikliu-analize?indicator=S2R003.
33. UNESCO. *Retrieved from*; 1994 https://whc.unesco.org/en/list/541/documents/. https://whc.unesco.org. https://whc.unesco.org/archive/repcom94.htm#541.
34. Husqvarna Urban Green Space Index. *Retrieved November 7, 2021, from*; 2020 https://www.hugsi.green/city?Vilnius. https://www.hugsi.green. https://www.hugsi.green/city?Vilnius.
35. Pinto LV, Ferreira CSS, Inácio M, Pereira P. Urban green spaces accessibility in two European cities: Vilnius (Lithuania) and Coimbra (Portugal). *Geogr Sustain* 2022;**3**:74–84. https://doi.org/10.1016/j.geosus.2022.03.001.
36. Vilnius municipality. *Vilnius master plan. Vilnius, Lithuania*; 2022. Retrieved January 21, 2022, from https://atviras.vplanas.lt/BP1/.
37. Marquart A, Eldridge DJ, Geissler K, Lobas C, Blaum N. Interconnected effects of shrubs, invertebrate-derived macropores and soil texture on water infiltration in a semi-arid savanna rangeland. *Land Degrad Dev* 2020;**31**:2307–18. https://doi.org/10.1002/ldr.3598.
38. Sun D, Zhang W, Lin Y, Liu Z, Shen W, Zhou L, et al. Soil erosion and water retention varies with plantation type and age. *For Ecol Manag* 2018;**422**:1–10. https://doi.org/10.1016/j.foreco.2018.03.048.
39. Dai L, Yuan Y, Guo X, Du Y, Ke X, Zhang F, et al. Soil water retention in alpine meadows under different degradation stages on the northeastern Qinghai-Tibet Plateau. *J Hydrol* 2020;**590**:125397. https://doi.org/10.1016/j.jhydrol.2020.125397.
40. Tehrany MS, Jones S, Shabani F. Identifying the essential flood conditioning factors for flood prone area mapping using machine learning techniques. *Catena* 2019;**175**:174–92. https://doi.org/10.1016/j.catena.2018.12.011.

41. Sørensen R, Zinko U, Seibert J. On the calculation of the topographic wetness index: evaluation of different methods based on field observations. *Hydrol Earth Syst Sci* 2006;**10**:101–12. https://doi.org/10.5194/hess-10-101-2006.
42. Eini M, Kaboli HS, Rashidian M, Hedayat H. Hazard and vulnerability in urban flood risk mapping: Machine learning techniques and considering the role of urban districts. *Int J Disaster Risk Reduct* 2020;**50**:101687. https://doi.org/10.1016/j.ijdrr.2020.101687.
43. Paliaga G, Faccini F, Luino F, Roccati A, Turconi L. A clustering classification of catchment anthropogenic modification and relationships with floods. *Sci Total Environ* 2020;**740**:139915. https://doi.org/10.1016/j.scitotenv.2020.139915.
44. Gomez RS, Garcia CB, Perez JC. The multicoll package versus other existing packages in R to detect multicollinearity. *Comput Econ* 2022;**60**:439–50. https://doi.org/10.1007/s10614-021-10154-1.
45. Ferdous MR, Di Baldassarre G, Brandimarte L, Wesselink A. The interplay between structural flood protection, population density, and flood mortality along the Jamuna River, Bangladesh. *Reg Environ Change* 2020;**20**:5. https://doi.org/10.1007/s10113-020-01600-1.
46. Huete AR. A soil-adjusted vegetation index (SAVI). *Remote Sens Environ* 1988;**25**: 295–309. https://doi.org/10.1016/0034-4257(88)90106-X.
47. Somvanshi SS, Kumari M. Comparative analysis of different vegetation indices with respect to atmospheric particulate pollution using sentinel data. *Appl Comput Geosci* 2020;**7**:100032. https://doi.org/10.1016/j.acags.2020.100032.
48. Li C, Liu M, Hu Y, Zhou R, Wu W, Huang N. Evaluating the runoff storage supply-demand structure of green infrastructure for urban flood management. *J Clean Prod* 2021;**280**:124420. https://doi.org/10.1016/j.jclepro.2020.124420.
49. Chen W, Wang W, Huang G, Wang Z, Lai C, Yang Z. The capacity of grey infrastructure in urban flood management: a comprehensive analysis of grey infrastructure and the green-grey approach. *Int J Disaster Risk Reduct* 2021;**54**:102045. https://doi.org/10.1016/j.ijdrr.2021.102045.
50. Mei C, Liu J, Wang H, Yang Z, Ding X, Shao W. Integrated assessments of green infrastructure for flood mitigation to support robust decision-making for sponge city construction in an urbanized watershed. *Sci Total Environ* 2018;**639**:1394–407. https://doi.org/10.1016/j.scitotenv.2018.05.199.
51. Maragno D, Gaglio M, Robbi M, Appioti F, Fano EA, Gissi E. Fine-scale analysis of urban flooding reduction from green infrastructure: an ecosystem services approach for the management of water flows. *Ecol Model* 2018;**386**:1–10. https://doi.org/10.1016/j.ecolmodel.2018.08.002.
52. Julia PB, Ferreira TM. From single- to multi-hazard vulnerability and risk in Historic Urban Areas: a literature review. *Nat Hazards* 2021;**108**:93–128. https://doi.org/10.1007/s11069-021-04734-5.
53. Schlumberger J, Ferrarin C, Jonkman SN, Diaz Loaiza MA, Antonini A, Fatorić S. Developing a framework for the assessment of current and future flood risk in Venice, Italy. *Nat Hazards Earth Syst Sci* 2022;**22**:2381–400. https://doi.org/10.5194/nhess-22-2381-2022.
54. Martínez-Gomariz E, Locatelli L, Guerrero M, Russo B, Martínez M. Socio-economic potential impacts due to urban pluvial floods in Badalona (Spain) in a context of climate change. *Water* 2019;**11**:2658. https://doi.org/10.3390/w11122658.
55. Kim M, Song K, Chon J. Key coastal landscape patterns for reducing flood vulnerability. *Sci Total Environ* 2021;**759**:143454. https://doi.org/10.1016/j.scitotenv.2020.143454.
56. Devi NN, Sridharan B, Kuiry SN. Impact of urban sprawl on future flooding in Chennai city, India. *J Hydrol* 2019;**574**:486–96. https://doi.org/10.1016/j.jhydrol.2019.04.041.

57. Manawi SMA, Nasir KAM, Shiru MS, Hotaki SF, Sediqi MS. Urban flooding in the northern part of Kabul City: causes and mitigation. *Earth Syst Environ* 2020;**4**: 599–610. https://doi.org/10.1007/s41748-020-00165-7.
58. Inacio M, Gomes E, Bogdzevič K, Kalinauskas M, Zhao W, Pereira P. Mapping and assessing coastal recreation cultural ecosystem services supply, flow, and demand in Lithuania. *J Environ Manag* 2022;**323**:116175. https://doi.org/10.1016/j.jenvman.2022.116175.
59. Gomes E, Inacio M, Bogdzevič K, Kalinauskas M, Karnauskaitė D, Pereira P. Future scenarios impact on land use change and habitat quality in Lithuania. *Environ Res* 2021;**197**:111101. https://doi.org/10.1016/j.envres.2021.111101.
60. Inacio M, Mikša K, Kalinauskas M, Pereira P. Mapping wild seafood potential, supply, flow and demand in Lithuania. *Sci Total Environ* 2020;**718**:137356. https://doi.org/10.1016/j.scitotenv.2020.137356.
61. Baro F, Hasse D, Gómez-Baggethun E, Frantzeskaki N. Mismatches between ecosystem services supply and demand in urban areas: A quantitative assessment in five European cities. *Ecol Indic* 2015;**55**:146–58. https://doi.org/10.1016/j.ecolind.2015.03.013.
62. Chen J, Jiang B, Bai Y, Xu X, Altalo JM. Quantifying ecosystem services supply and demand shortfalls and mismatches for management optimization. *Sci Total Environ* 2019;**650**:1426–39.
63. Javelle P, Fouchier C, Arnaud P, Lavabre J. Flash flood warning at ungauged locations using radar rainfall and antecedent soil moisture estimations. *J Hydrol* 2010;**394**: 267–74. https://doi.org/10.1016/j.jhydrol.2010.03.032.
64. Zhai X, Guo L, Zhang Y. Rainfall threshold determination for flash flood warning in mountainous catchments with consideration of antecedent soil moisture and rainfall pattern. *Nat Hazards* 2018;**94**:605–25. https://doi.org/10.1007/s11069-018-3404-y.
65. Salomao FS, Silva AM. Predicting soil erosion and assessing quality indicators in two Brazilian watersheds: subsidy for territorial planning. *Geocarto International* 2022;**37**: 4932–51. https://doi.org/10.1080/10106049.2021.1899308.
66. Jiang L, Zhu J, Wang H, Fu Q, Hu H, Huag Q. Spatial variability of the molecular composition of humic acids from subtropical forest soils. *J Soils Sediments* 2021;**21**: 766–74. https://doi.org/10.1007/s11368-020-02862-2.
67. Zhang Z, Yin H, Chang J, Xue J. Spatial variability of surface soil water content and its influencing factors on shady and sunny slopes of an alpine meadow on the Qinghai–Tibetan Plateau. *Glob Ecol Conserv* 2022;**34**:e02035. https://doi.org/10.1016/j.gecco.2022.e02035.
68. Kranz CN, McLaughlin RA, Johnson A, Miller G, Heitman JL. The effects of compost incorporation on soil physical properties in urban soils—a concise review. *J Environ Manag* 2020;**261**:110209. https://doi.org/10.1016/j.jenvman.2020.110209.
69. Pereira P, Ferreira A, Pariente S, Cerda A, Walsh RPD, Keesstra S. Preface: urban soils and sediments. *J Soils Sediments* 2016;**16**:2493–9. https://doi.org/10.1007/s11368-016-1566-3.
70. Hersperger AM, Oliveira E, Pagliarin S, Palka G, Verburg P, Bolliger J, et al. Urban land-use change: the role of strategic spatial planning. *Glob Environ Chang* 2018;**51**:32–42. https://doi.org/10.1016/j.gloenvcha.2018.05.001.
71. Bidandi F, Williams JJ. Understanding urban land, politics, and planning: a critical appraisal of Kampala's urban sprawl. *Cities* 2020;**106**:102858. https://doi.org/10.1016/j.cities.2020.102858.
72. Hansen R, Olafsson AS, van der Jagt AN, Rall E, Pauleit S. Planning multifunctional green infrastructure for compact cities: what is the state of practice? *Ecol Indic* 2019;**96**:99–110. https://doi.org/10.1016/j.ecolind.2017.09.042.

73. Morelli AB, Cunha AL. Measuring urban road network vulnerability to extreme events: an application for urban floods. *Transp Res Part D: Transp Environ* 2021;**93**:102770. https://doi.org/10.1016/j.trd.2021.102770.
74. Kalinauskas M, Bogdzevič K, Gomes E, Inácio M, Barcelo D, Zhao W, et al. Mapping and assessment of recreational cultural ecosystem services supply and demand in Vilnius (Lithuania). *Sci Total Environ* 2022;**855**:158590. https://doi.org/10.1016/j.scitotenv.2022.158590.

CHAPTER SEVEN

Mapping of ecosystem services: Supply and demand for local climate regulation and nutrient regulation services

Zahra Kalantari[a,b,*], Samaneh Seifollahi-Aghmiuni[b], Farzad Vahidi Mayamey[b,c], Navid Ghajarnia[b], and Riad Lilipaly[b]

[a]Department of Sustainable Development, Environmental Science and Engineering (SEED), KTH Royal Institute of Technology, Stockholm, Sweden

[b]Department of Physical Geography and Bolin Centre for Climate Research, Stockholm University, Stockholm, Sweden

[c]Department of Civil, Chemical, Environmental, and Materials Engineering (DICAM), University of Bologna, Bologna, Italy

[*]Corresponding author: e-mail address: zahrak@kth.se

Contents

Abstract

Global population growth and associated socio-economic development have led to rapid urban expansion worldwide, with management implications for sustainable natural resources and societal resilience. Natural ecosystems and the services they provide are essential for societal mitigation and adaptation to adverse environmental consequences in urban areas. Mapping ecosystem services is a valuable tool in spatial

ISSN 2468-9289
https://doi.org/10.1016/bs.apmp.2022.09.002

planning for urban development, as it provides a deeper understanding of complex human-natural system interactions. This study analyzed and mapped two ecosystem services (local climate regulation and nutrient regulation), which play a key role in mitigating the impacts of local and global climate change in urban areas and of nutrient loads entering surface waters. The specific cases analyzed (Amsterdam city and the Netherlands as a whole) provided insights into opportunity pathways for adaptive development and management of complex urban environments and can support policy and decision-making processes for a sustainable and resilient future.

Keywords: Urban areas, Urban management, Land spatial planning, Climate change adaptation, Mitigation

1. Introduction

The global population continues to grow, albeit at a slower pace than at any time since 1950. From an estimated level of 7.7 billion people worldwide in 2019, projections indicate population growth to around 8.5 billion by 2030, 9.7 billion by 2050, and over 11 billion by 2100.[1] At present, around 54.9% of the global population lives in urban areas and this proportion is expected to reach 66% by 2050.[2] This increase in the urban share of the total population, i.e., the process of urbanization, affects natural resources in urban areas and the services they provide.

Urban ecosystems can be considered to comprise all green and blue areas in cities, including street trees and ponds. These ecosystems support air purification, carbon sequestration and water regulation, and increase the resilience of cities to extreme weather events such as heatwaves and floods.[3] However, the capacity of ecosystems to supply particular services is intimately tied to land cover change, which affects the ability of ecosystems to deliver sufficient services.[4] Ecosystems also differ from one another in their structures and processes, and in their capacity to supply a specific service.[5] Natural factors (climate, hydrology) and human influences (climate change, land use change) can have an impact on ecosystem services supply, i.e., the capacity of a given region to provide ecosystem services during a given time. Demand for ecosystem services is represented by the total of all ecosystem products and services currently consumed or used over a given period in a certain location.[5]

Natural ecosystem services play a vital role in addressing urban challenges by improving air quality and promoting the wellbeing and health of urban residents.[6] Man-made ecosystems, including villages, towns, cities, rivers, orchards, dams, gardens, lakes and agriculture, can also enhance social-ecological values and promote social inclusion by providing public space for community gathering and leisure activities in cities.[7,8] However, humans generally alter land cover and ecosystems to fulfil their basic needs.[9,10] To accommodate the growing global demand for food, housing and industrial goods, natural ecosystems are being converted to cropland and built-up land. According to the Millennium Ecosystem Assessment (MEA), around 60% of ecosystem services were degraded globally in the latter half of the 20th century.[11] The stresses on ecosystems imposed by climate change and human activities have affected ecosystem function and reduced the various benefits ecosystems provide to human societies, through biodiversity loss, soil degradation and, in some cases, irreversible damage.[12,13]

Land consumption as a result of urbanization causes surface sealing, soil compaction and soil erosion and can restrict the ability of land to supply ecosystem services.[14] It also increases the susceptibility of urban areas to natural disasters and climate change impacts such as flood, drought and heatwaves, with implications for water, energy and food supply.[15] Urbanization can also affect water resources by (i) replacing vegetation with sealed surfaces, (ii) altering soil physical properties such as aggregate size, porosity and soil hydraulic properties, and (iii) using artificial infrastructure such as drainage systems to collect excess runoff from urban surfaces. This can undermine services that could potentially be provided by local ecosystems (e.g., water purification, water retention, climate regulation) at a low cost with high adaptability and several co-benefits.[16]

The MEA divides ecosystem services into four categories (supporting, provisioning, regulating, cultural).[11] Mapping and analyzing different categories of ecosystem services can support sustainable utilization and management of land and water resources through long-term spatial planning and urban management.[17] Ecosystem service mapping has also been identified as a critical component in improving recognition of ecosystem services and their integration into institutional and decision-making processes.[18]

The concept of ecosystem services in urban built environments is introduced below. In case studies described and reported in subsequent sections, two ecosystem services in the Netherlands are analyzed and mapped. The two cases are: (i) Local climate regulation service in Amsterdam metropolitan area; and (ii) nutrient regulation service in the whole of the Netherlands. These two regulating services are evaluated and quantified based on a method introduced by Burkhard et al.[5] and refined by Goldenberg et al.[19]

2. Ecosystem services in urban environments

Among the four main categories of ecosystem services (i.e., supporting, provisioning, regulating and cultural), *supporting services*, such as soil formation, photosynthesis and nutrient and water cycles, often do not have direct impacts on humans, animals and plants. However, they are an essential prerequisite for provision of other types of ecosystem services. *Provisioning services* are the (physical) products obtained from ecosystems, such as food, fiber, fuel and water. These services are commonly best known and best understood by the public when discussing ecosystem services. *Regulating services* are the advantages derived from regulation of ecosystem processes, such as air quality regulation, climate regulation, water regulation and purification, and disease and pest regulation. *Cultural services* are the ethereal benefits people obtain from ecosystems, such as spiritual and religious values, educational values, cultural heritage values and recreation.[11]

Urban ecosystem services are provided in a variety of environments, including: *green areas* such as parks, woods and cemeteries, *open spaces* such as squares, pedestrian areas, cycling areas, gardens, yards and campus areas, and *blue spaces*, including streams, lakes, ponds, artificial swales and stormwater ponds. Soil and water regulation, local climate regulation and nutrient regulation are among various functions provided by these ecosystems, e.g., urban green zones can regulate local temperature, store water and reduce runoff during extreme weather events.[20] Compared with less densely populated rural areas, urban areas have high demand/use for ecosystem services.[12] To better understand and map ecosystem services in urban areas, it is crucial to link their supply and demand, i.e., map the pattern of service realization.

Apart from the ecosystems located inside cities and urban areas, other ecosystems outside the urban boundaries, e.g., lakes and forests, can also play a role in providing regional services, as the whole natural system is ultimately interconnected. Lakes can provide services such as water supply and water quality. Forests can help control erosion[21] and reduce runoff as a result of increased capacity for infiltration and storage. This can improve the flood resilience of cities, especially those located downstream in river catchments.[22,23] However, research has shown that soil condition is the most important factor linking land use and flood regulation, and that much of the soil degradation associated with deforestation is caused by poor land-use practices (e.g., road and building construction).[24]

Local climate regulation is an important urban ecosystem service, providing a significant benefit to local citizens, specifically through reducing the urban heat island (UHI) effect.[25] Parks, woods, lawns, gardens, rivers, lakes and ponds regulate the temperature. This cooling effect is influenced by the size of the green area, although it is unclear whether there is a minimum size requirement or whether these two parameters have any direct relationship.[26] As a result, evaluation and management of urban structures is necessary to adapt metropolitan areas to climate change impacts, such as heat waves,[27] which are happening more frequently around the world and in Europe, most recently in 2003, 2010, 2018 and 2020[28] and 2022.

Nutrient regulation service is the ability of an ecosystem to regulate nutrients on land and in water urbanization and other human activities directly manipulate the nutrient cycle (mainly nitrogen and phosphorus), one of the numerous environmental processes that natural ecosystems regulate.[14] Plant growth is influenced by nutrient availability. Excess nutrients that are not used by plants typically accumulate in soils, sediments, subsurface flow and surface water, where build-up of large amounts can result in eutrophication, i.e., enrichment of inland and coastal waters with anthropogenically-driven inputs of nitrogen and phosphorus, and associated water quality issues.[29]

3. Case study areas

Two case study areas, the city of Amsterdam and the whole of the Netherlands, were used as cases for mapping the two selected regulation services (local climate, nutrient). The Netherlands (Fig. 1A) has a temperate

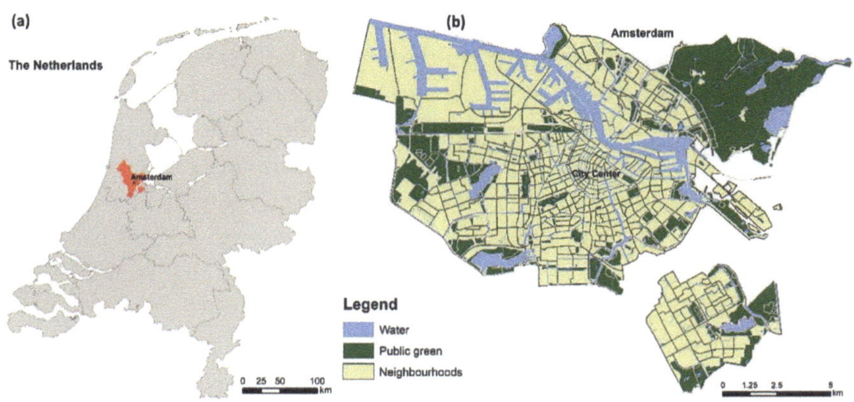

Fig. 1 (A) Map of the Netherlands indicating the location of the capital city, Amsterdam and (B) map of the Amsterdam metropolitan area.

maritime climate, with mild summers and winters.[30] Mean annual temperature at national level is 10.4 °C and mean annual precipitation is 790.9 mm (data for 1991–2020).[31] Land cover in the Netherlands consists of agricultural areas (57%), open water (18%), forest and meadows (12%), urban areas (5%), infrastructure (4%) and recreational areas (2%).[32] These values differ slightly from the CORINE land cover data published in 2018, based on which agricultural land in the Netherlands makes up 51%, 22% is covered by open water, 9% is under forest and natural grasslands, about 12% is covered by artificial areas (urban, infrastructure and sports facilities) and marshes and intertidal flats make up the remaining 6%.[13]

During the first half of the 20th century, many forests and grasslands in the Netherlands were converted to agricultural land. In subsequent decades, arable land and meadows were further transformed for the development of urban and industrial areas.[33] These changes in land use affected water management in the Netherlands, which is a dynamic process with shallow groundwater levels and a strong interaction between groundwater and surface water.[34] Amsterdam (Fig. 1B) is the capital and largest city in the Netherlands, with more than 870,000 inhabitants in 2021.[35] It acts as a great example of how water has influenced urban design, formed and developed in relation to river and sea. The city occupies an area of 219 km^2, of which 166 km^2 is covered by land and the remainder by water.[35]

4. Mapping ecosystem services in the study areas

4.1 Local climate regulation

Methodology introduced by Burkhard et al.[5] Burkhard and Maes[36] and refined by Goldenberg et al.[19] was used for assessing and quantifying potential supply and demand for the two selected ecosystem services: local climate regulation (LCR) in the city of Amsterdam and nutrient regulation across the Netherlands. The LCR service was only assessed for Amsterdam because demand for this service arises mainly in metropolitan areas. The methodology used is based on a non-monetary assessment involving various indicators that are classified and mapped with respect to relative supply and demand. The overall outcome is an estimate of the capacity for providing a particular ecosystem service, based on landscape characteristics defining land use/land cover conditions.

Urban Atlas (UA) land cover data from Copernicus Land Monitoring Service[13] were used to assess and map supply and demand of the LCR service in Amsterdam. Urban Atlas provides land use and land cover data for different European cities, categorized in 27 different land cover classes (17 artificial surfaces, five agricultural areas, three semi-natural areas and two wetland-water categories). The data are based on visual interpretation of very high-resolution satellite imagery and the dataset utilizes a minimum mapping unit (MMU) of 0.25 ha in urban areas and 1 ha in rural areas (Copernicus Land Monitoring Service[13]).

To assess potential supply and demand for the LCR service in Amsterdam, different landscape classes defined in the UA land cover data were assigned an indicator value within the range 0 (no demand for a specific ecosystem service and/or no relevant capacity to provide that service) to 5 (very high demand for a specific ecosystem service and/or very high relevant capacity to provide that service). Note that the 0–5 scale is relative and does not specifically quantify the exact supply capacity (or demand) for a specific ecosystem service. Assigning the same indicator value to different land cover classes does not necessarily mean that they act similarly regarding a specific ecosystem service or they have the same capacity to provide a specific ecosystem service.

Since the land cover classes defined in the UA data might not necessarily match actual land cover classes in the study area, indicator values were defined based on Burkhard et al.[5] when there was a direct match between the two land cover classes. By combining landscape data and indicator values, easy-to-read maps were produced showing: (1) natural conditions and human activities; and (2) the potential supply and demand of different ecosystem services. An assessment matrix was then created to link the different UA land cover classes (listed in rows in Table 1) to supply and demand of the LCR service (listed in columns).

In the assessment matrix, each land cover class was given two indicator values, for supply and demand in relation to the LCR service. Hypothetically, if an area can provide a specific ecosystem service and at the same time requires a similar amount of that ecosystem service, service supply will be realized in that area and supply will meet demand. For instance, different urban area classes are defined in the matrix by their content of artificial structures and average degree of soil sealing (S.L.) (i.e., number of buildings or any impermeable artificial material that permanently covers the ground surface). The discontinuous low-density urban fabric class (S.L. 10–30%) has more urban green areas, parks, gardens etc. than the discontinuous dense urban fabric class (S.L. 50–80%), and thus theoretically has more capacity to provide various ecosystem services with less demand for a specific service. However, most classes are similar and thus indicator values

Table 1 Assessment matrix of indicator values of supply and demand for the local climate regulation (LCR) service in relation to different land cover types defined in Urban Atlas data.

Urban Atlas land cover type	Local Climate Regulation (LCR)	
	Demand	Supply
Continuous urban fabric (S.L.: > 80%)	5	0
Discontinuous dense urban fabric (S.L.: 50% - 80%)	5	0
Discontinuous medium density urban fabric (S.L.: 30% - 50%)	4	1
Discontinuous low-density urban fabric (S.L.: 10% - 30%)	3	2
Discontinuous very low-density urban fabric (S.L.: < 10%)	2	3
Isolated structures	1	3
Industrial, commercial, public, military and private units	1	0
Fast transit roads and associated land	2	0
Other roads and associated land	2	0
Railways and associated land	2	0
Port areas	2	0
Airports	2	0
Mineral extraction and dump sites	0	0
Construction sites	2	0
Land without current use	1	2
Green urban areas	2	2
Sports and leisure facilities	2	1
Arable land (annual crops)	2	3
Permanent crops	2	3
Pastures	1	1
Forests	0	5
Herbaceous vegetation associations	0	2
Open spaces with little or no vegetation	0	0
Wetlands	0	4
Water	0	2

Demand	
0	No relevant demand
1	Low relevant demand
2	Relevant demand
3	Medium relevant demand
4	High relevant demand
5	Very high relevant demand

Supply	
0	No relevant capacity
1	Low relevant capacity
2	Relevant capacity
3	Medium relevant capacity
4	High relevant capacity
5	Very high relevant capacity

for the study area of Amsterdam were taken directly from the original assessment matrix developed and proposed by Burkhard et al.[5] To map the LCR service in the city of Amsterdam, the indicator values defined in Table 1 were applied to their respective land cover classes, using ArcMap 10.6.1. This was done for both potential supply and potential demand for this LCR service.

In the original methodology, the potential supply and demand values obtained for different land cover classes are merged to give the overall net budget for the service under study.[5,36] This reveals spatial patterns, with areas where the demand can/cannot be met and areas with potential supply or demand for this service. Goldenberg et al.[19] refined the method for calculating the net budget value of ecosystem services by fitting a layer of 150 m wide hexagons representing expected geographical range of local air movement (which redistributes heat and mixes local air temperatures, dispersing temperature anomalies within each spatial unit) on top of the potential supply and demand map (Fig. 2).

Based on this airflow-adjusted method, the net budget for the LCR service in the city of Amsterdam was calculated using the Tabulate Intersection tool in ArcMap. This tool cuts all the vector lines (bottom layer) that

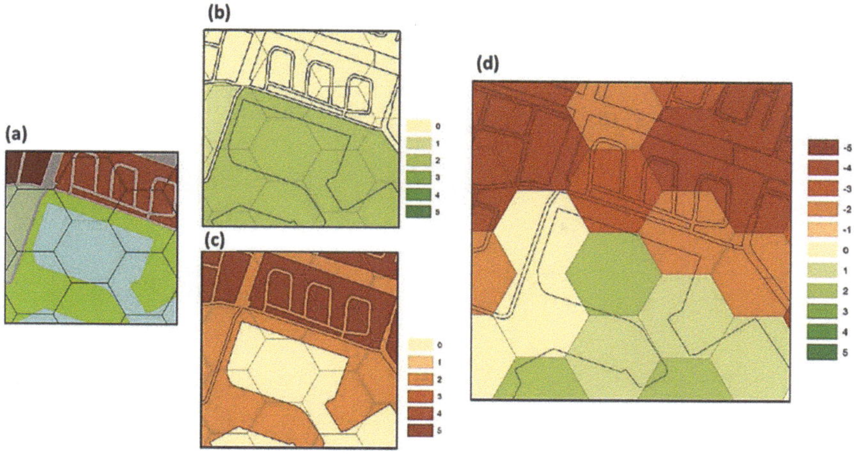

Fig. 2 Graphical example of net budget calculation for the local climate regulation (LCR) service. (A) Urban Atlas land cover map showing different land cover classes (see color key); (B) potential supply and demand map; (C) potential supply-demand map with a 150-m hexagon top layer fitted on top; and (D) map showing calculated net budget.

intersect with each hexagon (top layer). Each hexagon then consists of multiple smaller land cover units. The net budget value in one hexagon unit H is calculated as:

$$H = \frac{\sum_{i=1}^{n} A_i * S_i + \sum_{j=1}^{m} A_j * D_j}{A_H} \tag{1}$$

where A_i and A_j represent area of the smaller land cover units; S_i is the score for potential supply of element i; D_j is the score for potential demand for element j; and A_H represents the area of one hexagon unit (14,614 m^2).

The net budget values obtained with this methodology range from -5 to $+5$, with negative (positive) values indicating high demand (supply) for a specific ecosystem service. The zero (neutral) value indicates realization of both potential supply and potential demand for a specific ecosystem service (Fig. 3).

4.2 Nutrient regulation

For the NR service, potential supply can be assessed and quantified following a similar approach as explained for the LCR service above. However, since other parameters and factors are involved in demand for the NR service, demand mapping is performed slightly differently than for the

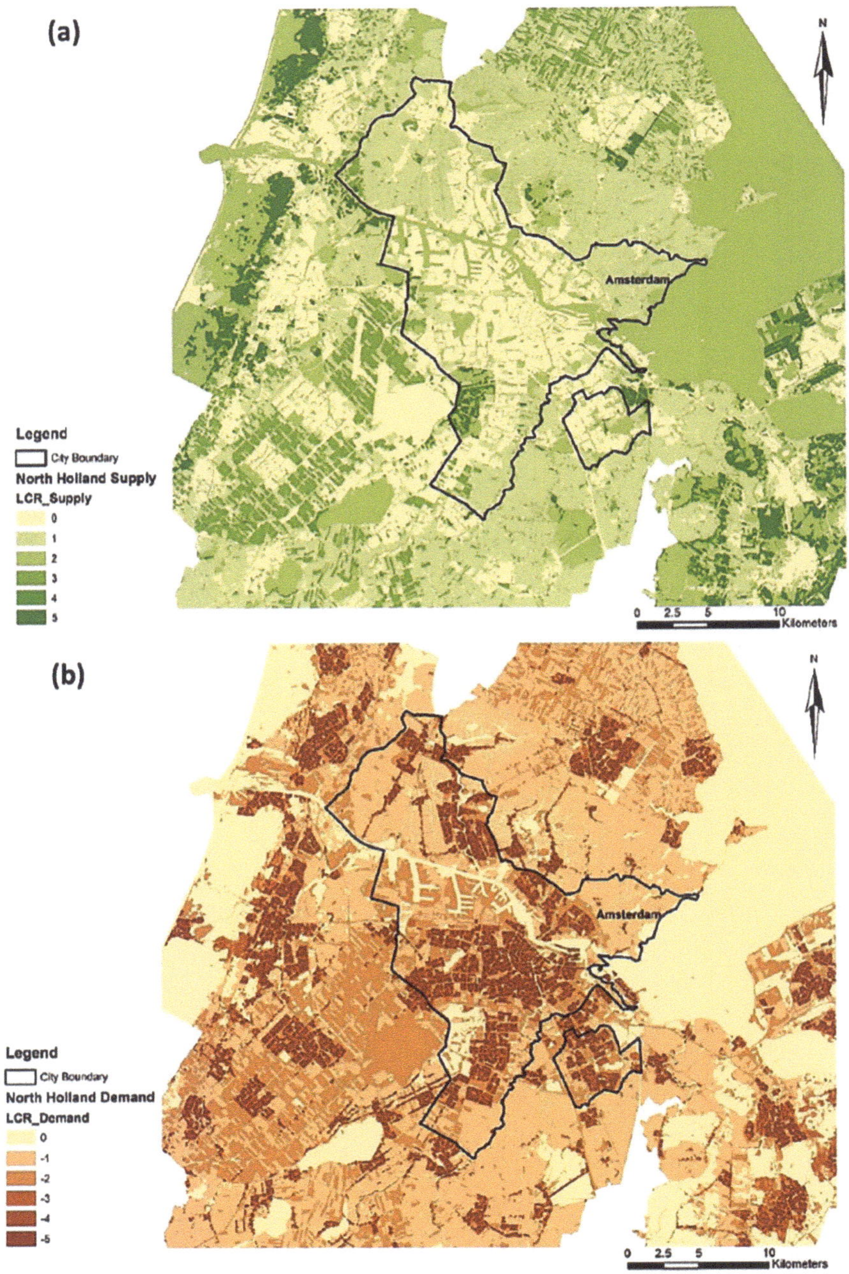

Fig. 3 (A) Potential supply and (B) potential demand maps for the local climate regulation (LCR) service in the city of Amsterdam and surrounding urban areas.

LCR service. In the present case, CORINE land cover data from 2018 were used to assess supply (and demand) for the NR service across the Netherlands (Table 2). This dataset utilizes a MMU of 25 ha for the whole of the Netherlands and consists of 44 classes categorized in five groups: artificial surfaces, agricultural areas, forest and semi-natural areas, wetlands, and water bodies. In addition to the CORINE land cover data, in NR service mapping other types of data, including shapefiles of 538 hydrological catchments across the Netherlands, annual phosphorus and nitrogen loads from different sectors and their retention rate in around 300 catchments were used to assess potential demand.

Table 2 Assessment matrix of indicator values of supply and demand for the nutrient regulation (NR) service in relation to different land cover types defined in the CORINE data.

CORINE land cover type	Nutrient Regulation (NR)	
	Demand	Supply
Continuous urban fabric (S.L.: > 80%)	1	0
Discontinuous dense urban fabric (S.L.: 50% - 80%)	2	0
Industrial, commercial, public, military and private units	3	0
Fast transit roads and associated land	0	0
Other roads and associated land	0	0
Railways and associated land	0	0
Port areas	0	0
Airports	1	0
Mineral extraction and dump sites	0	0
Dump Sites	0	0
Construction sites	2	0
Green urban areas	0	2
Sports and leisure facilities	0	1
Arable land (annual crops)	3	1
Permanent crops	5	1
Fruit trees and berries	3	2
Pastures	1	1
Complex cultivation patterns	5	1
Agriculture and Natural vegetation	3	2
Agro-foresty areas	3	2
Broad-leaved forest	0	5
Coniferous forest	0	5
Mixed forest	0	5
Moors and Heathland	0	3
Sclerophyllous vegetation	0	2
Transitional woodland shrub	0	2
Beaches, dunes and sand plains	0	1
Bare rock	0	0
Sparsely vegetated areas	0	1
Burnt areas	0	0
Glaciers and perpetual snow	0	0
Inland marshes	0	4
Peatbogs	0	4
Salt Marshes	0	2
Salines	0	0
Intertidal flats	0	1
Wetlands	0	3
Water	0	3
Coastal Lagoons	0	3
Estuaries, Sea and Ocean	0	3

Demand	
0	No relevant demand
1	Low relevant demand
2	Relevant demand
3	Medium relevant demand
4	High relevant demand
5	Very high relevant demand

Supply	
0	No relevant capacity
1	Low relevant capacity
2	Relevant capacity
3	Medium relevant capacity
4	High relevant capacity
5	Very high relevant capacity

The majority of excess nutrients in a landscape originate from agricultural activities, water/wastewater/stormwater treatment plants and atmospheric deposition. Therefore it is crucial to identify the location of these sources across the country, as those areas have a stronger need for the NR service. Here, a method by Bicking et al.[37] was used to map potential demand for the NR service in different parts of the Netherlands. For this, the nutrient budget was calculated based on the total nutrient loads from atmospheric deposition, treatment plants, industrial sites, agricultural areas, soil leaching, inflow waters and other sources, using data retrieved from the Pollutant Release and Transfer Register's online portal (http://www.emissieregistratie.nl/). Data for agricultural activities were obtained using a nutrient emission modelling system called STONE (the Dutch acronym for Jointly Developed Nutrient Emission Model).[38]

The calculated nutrient budget can be mapped in connection with different land cover classes, showing the amounts of nutrient loads entering a recipient system (e.g., a water body) without reflecting the quality changes in that system. A hydrological catchment that receives high (low) nutrient loads can still have good (bad) ecosystem (water) quality, owing to provision (lack of) the NR service. The nutrient loads entering recipient bodies in each catchment are retained during transport (e.g., in water) pathways or used by terrestrial and aquatic plants. Nutrients also can escape in gaseous form into the atmosphere as a result of denitrification. The amount of nutrient uptake across a catchment is known as nutrient retention. Multiplying the calculated nutrient budget by the retention rate in each catchment gives the total amount of nutrients captured/retained in that catchment. Accordingly, the residual nutrient loads in each catchment can be mapped as an indicator of areas in need of additional management measures to lower nutrient concentrations (i.e., potential demand for the NR service).

4.3 Local climate regulation (LCR) in the city of Amsterdam

The assessment matrix for the LCR service indicated demand for this service for almost all the land cover classes in Amsterdam (Table 1). For instance, the demand in arable areas and pasture areas was assigned an indicator value of 2 and 1, respectively, because plants growing on cropland and animals grazing on pastures also need the LCR service. Continuous urban fabric (S.L. >80%) and discontinuous dense urban fabric (S.L. 50–80%) areas had the highest demand for the LCR service and these areas also cannot, or have low potential capacity, to provide this service. Natural land cover types (green and blue areas such as forests and wetlands) are more capable of providing the LCR service, and were thus assigned the highest supply values (4, 5).

A net budget map for the LCR service in Amsterdam and surrounding urban areas was produced by merging the supply and demand maps for this service. The net budget values ranged from −5 (high demand) to +5 (high supply) (Fig. 4).

As can be seen in Fig. 4, the city itself and the many smaller cities outside the capital (e.g., Haarlem, Zaandam, Amstelveen, Purmerend) had the highest potential demand and lowest potential supply. In fact, potential demand for the LCR service far exceeded the potential supply within the city boundaries. Different water bodies (such as canals, lakes and the Het IJ river) and some small parks scattered in and around the city center were the only areas that could supply the LCR service. Small agricultural areas (located in the north, north-east, and far south within the city boundary) were also shown to be capable of providing this service, but since these areas have corresponding demand for this service, they ended up as neutral areas (with value 0) in the net budget map (Fig. 4). In both the potential supply map and the net budget map, Amsterdam forest, located to the south within the city boundary, appeared as the only area with sufficient relevant capacity

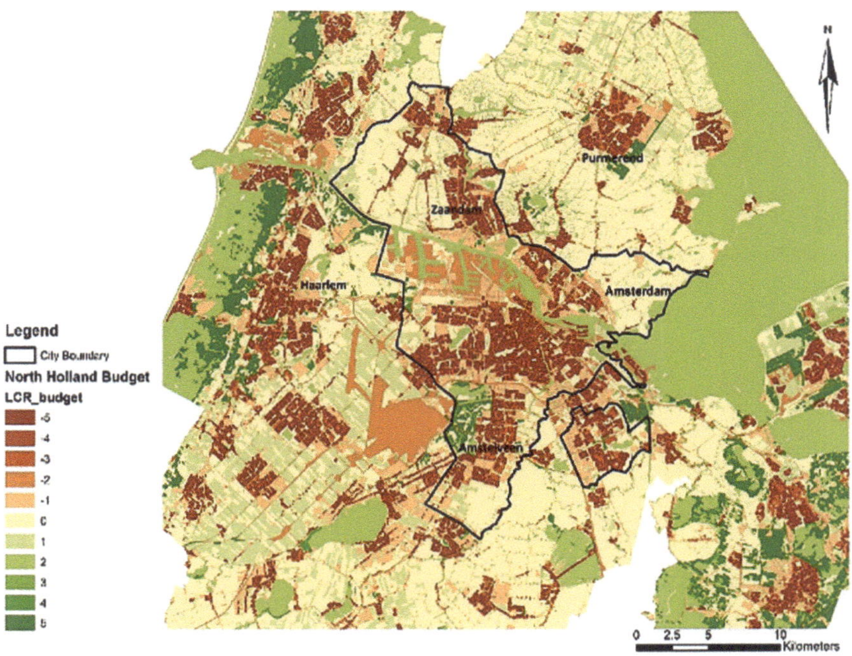

Fig. 4 Net budget for the local climate regulation (LCR) service in the city of Amsterdam and surrounding urban areas.

to provide the LCR service, with high potential supply and low potential demand (Fig. 3A and B). Zuid-Kennermerland National Park, located on the west coast, and Westerheide (small forest and heathland areas), located east of Amsterdam, also showed high supply capacity (Fig. 3A).

Fig. 5 shows the net budget estimated with the hexagon overlay method (considering local air movement) for the LCR service in the city of Amsterdam itself and in Zaandam to the north and Amstelveen to the south,

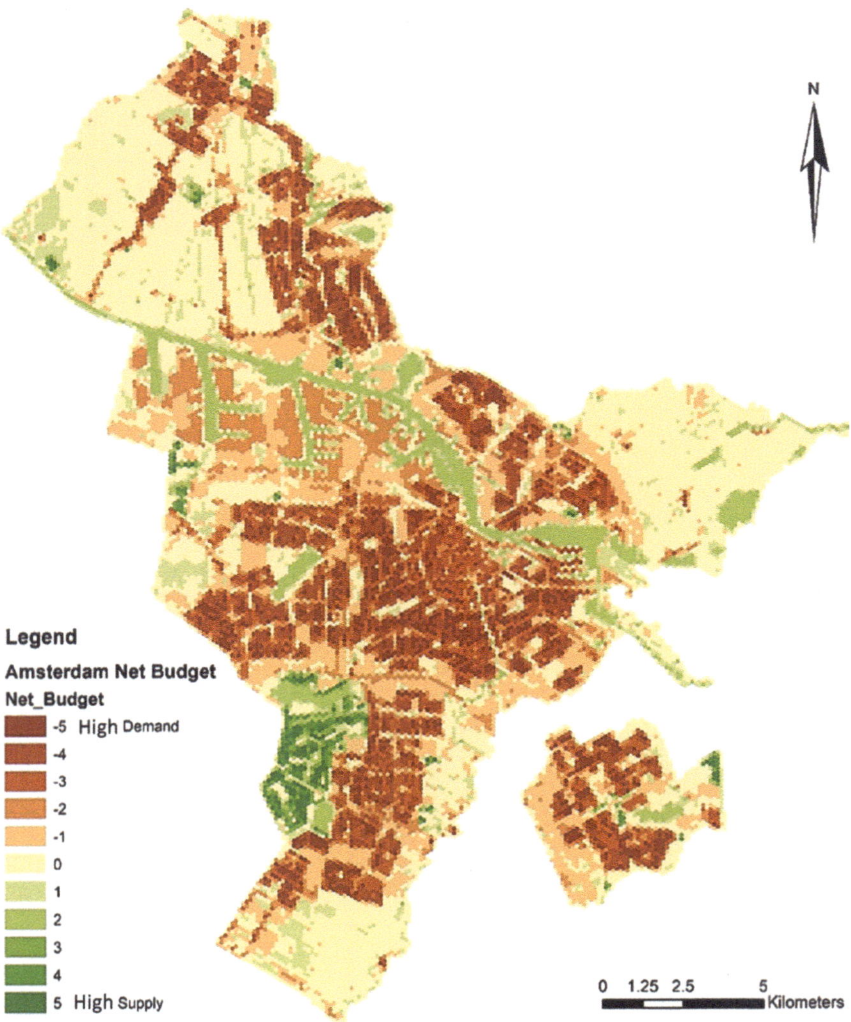

Fig. 5 Net budget for the local climate regulation (LCR) service in the city of Amsterdam and surrounding metropolitan areas as estimated using the hexagon overlay method (considering local airflow).

all densely populated metropolitan (urban) areas. All these areas showed substantial potential demand for the LCR ecosystem service according to this mapping approach. However, looking more closely at the city center, the net budget values ranged from −2 to −5 (Fig. 5), whereas the supply-demand map for the city and surrounding area only showed values of −5 (Fig. 4). This difference is because the hexagon overlay methodology considers the airflow that distributes the LCR service, linking demand and supply, and provides a better perspective on the service distribution.

Fig. 6 shows a close-up comparison of the net budget for the LCR service in Amsterdam based on the two mapping methods applied. Sloterplas (box 1) is a small lake located in western Amsterdam, adjacent to four different municipalities. The lake and its surrounding park are popular for recreation and different (water) sports and activities. The lake itself had a relatively high positive net budget for the LCR service, representing higher supply than demand for this service (indicated by light green in Fig. 6A), while the surrounding park was well-balanced with a more neutral net budget value (indicated by yellow in Fig. 6A). There were high negative net budget values for the LCR service in adjacent residential areas (Fig. 6B), representing higher demand than supply of this service (dark red in the map). The net budget values gradually increased toward positive values on moving toward lakes and parks (areas with neutral net budget or

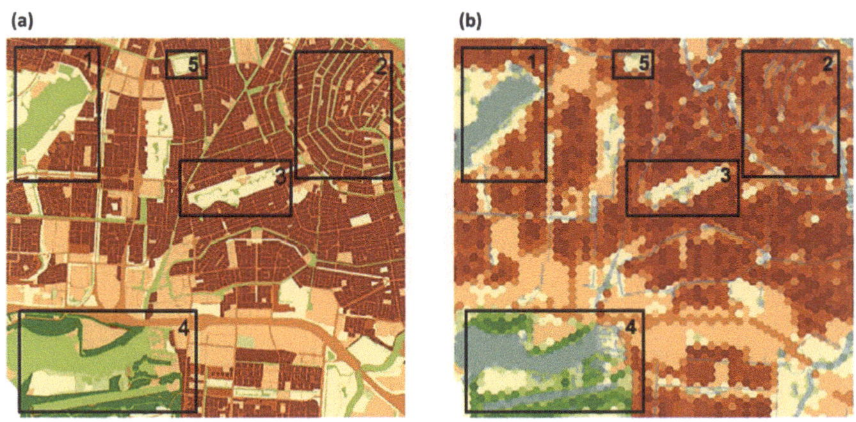

Fig. 6 Net budget for the local climate regulation (LCR) service in selected sub-areas in the city of Amsterdam: Sloterplas (box 1), Canal District (box 2), Vondelpark (box 3), New Lake and part of Amsterdam forest (box 4) and Erasmus Park (box 5). (A) Net budget map without averaging the score values and (B) net budget map with score averaging in hexagons. The blue water layer is added to map (B) for better readability.

higher supply of the LCR service). A similar pattern was found for the largest park in central Amsterdam, the 47-ha Vondelpark (box 3), which showed a well-balanced supply-demand budget (yellow in both maps in Fig. 6). The residential buildings closest to the park showed lower negative net budget values (lower potential demand; light red hexagons in Fig. 6B) than buildings farther away (higher potential demand than supply; dark red hexagons in Fig. 6B). The LCR mapping for Erasmus Park (box 5), north-west of Vondelpark, also revealed an increase in potential demand on moving farther away from the park. In both parks, small ponds provided the LCR service (positive net budget, representing potential supply). Many small local parks scattered between residential areas in different neighborhoods throughout Amsterdam were shown to provide the LCR service on small local scale, appearing as scattered yellow hexagons surrounded by red hexagons with high negative net budget values representing high potential demand for the LCR service (Fig. 6B).

The Canal District (box 2), in the city center, consists of four major canals that are connected to the Amstel and IJ rivers. The area is characterized by dense buildings built in the 17th century. Relatively small young trees grow along these canals and most of the urban green areas are private gardens located between the buildings. In LCR service mapping, all water bodies in the city were found to be characterized by positive net budget values (potential supply, green in Fig. 6A), with their surrounding built environments representing a negative net budget (potential demand, red in Fig. 6A). However, on using the hexagon method for mapping the LCR service, the Canal District area was characterized overall by lower negative net budget values (lower potential demand) (light red in Fig. 6B) compared with Fig. 6A (higher negative net budget values; more dark red areas). Along the water bodies and pathways in the district, hexagons associated with the closest areas were all characterized by lower negative net budget values (lower potential demand; light red in Fig. 6B) than those located far away. The neighborhoods outside the Canal District (north and south-east of Vondelpark) are heavily/densely built and the proportion of urban green areas is low (dark red in Fig. 6B).

New Lake and Amsterdam forest (box 4) are located just outside Amsterdam. There is a large industrial area, sport facilities, a highway and railway between the city and the lake. These areas were characterized by low negative net budget values (low demand for the LCR service, light red in Fig. 6). The forest is on the east of this area, bordering the municipality of Amstelveen and, on its west side, Schiphol airport. The forest was the area with the highest positive net budget (high potential supply) of the LCR service (Fig. 6). Amstelveen municipality appeared as an area with the whole range of negative net budget (high potential demand), possibly due to different types of housing built in this area, as some neighborhoods consist of densely-built high apartments with fewer urban green areas while others include single-family homes and more green spaces and private gardens. Amsterdam forest appeared to provide a considerable supply of the LCR service, but surrounding areas still had net potential demand for this service.

4.4 Nutrient regulation (NR) across the Netherlands

The assessment matrix for the NR service (Table 2) showed that agricultural land and agroforestry areas in the Netherlands require this service. Excessive nutrient uptake by crop plants can be hazardous and can lead to growth issues.[39] The natural processes of nutrient removal lie mainly in the denitrification process, plant uptake and sedimentation. Natural land cover types are more capable of providing the NR service, but their supply ability depends on various factors such as climate conditions, areas of open water, vegetation, soil type and soil texture.

Fig. 7 shows potential supply of the NR service, mapped based on the indicator values in Table 2 and their relationship with different land cover classes defined in the CORINE (2018) land cover data. As can be seen, the Netherlands tended to be characterized by low potential supply of the NR service (light green areas), but with small patches of no potential supply (pink) distributed all over the country, mostly associated with cities and urban areas. Small areas with high potential supply (dark green) were also distributed throughout the country, but particularly in the south, east and northeast. Veluwe forest, a national park in the center of the country that is one of the largest nature reserves (5500 ha),[40] showed the highest potential supply of the NR service (dark green in Fig. 7). Other areas with relatively high potential supply were identified along the coast. Western parts of the Netherlands, which are strongly dominated by agricultural land, cities and small villages, showed intermediate potential supply of the NR service.

Fig. 8 shows the annual nutrient (nitrogen and phosphorus) loads entering surface waters in the period of 2010–2013 for all studied catchments in the Netherlands. Relevant data were collected only for regional waters managed by local authorities. National waters, such as the major rivers

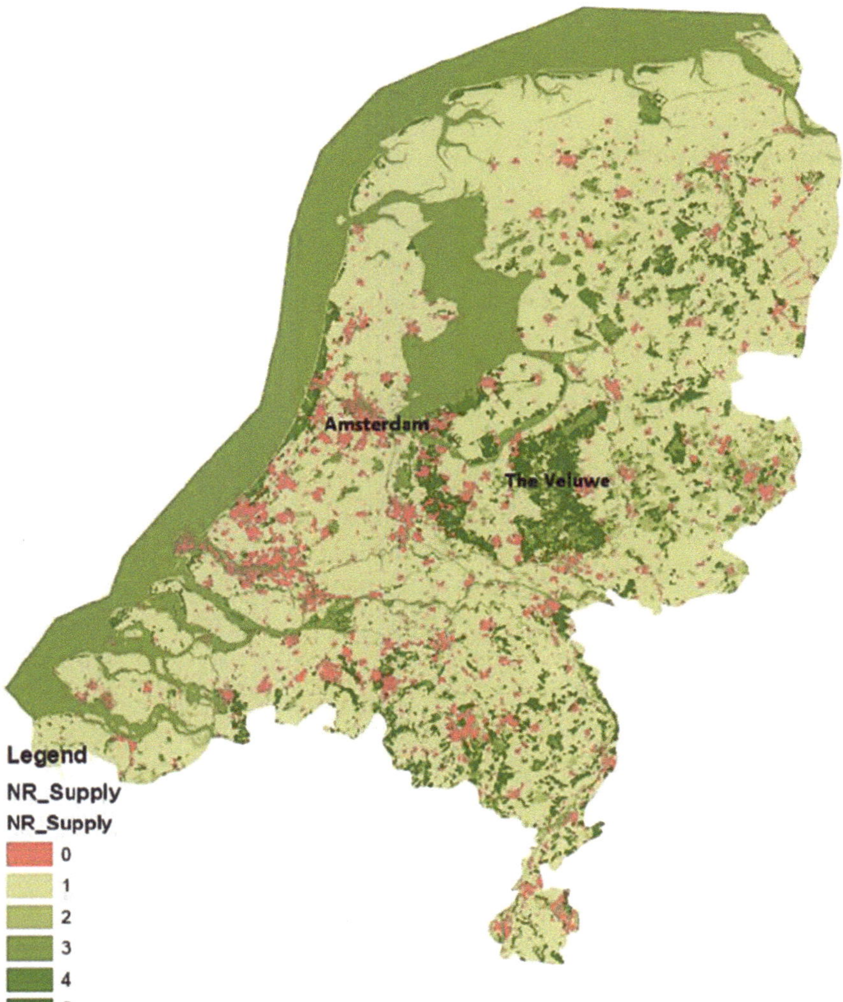

Fig. 7 Potential supply of the nutrient regulation (NR) service across the Netherlands.

(Rhine, Meuse, Scheldt) and coastal waters managed by central government were not considered, since 60–75% of their nutrient loads come from neighboring countries. As can be seen in Fig. 8, more than 80% of studied catchments across the country had low nitrogen loads entering their water bodies in the study period (less than 40 kg per hectare & year). High nitrogen loads to surface waters were seen in less than 10% of the catchments, distributed across the country. For phosphorus, almost all catchments showed low or

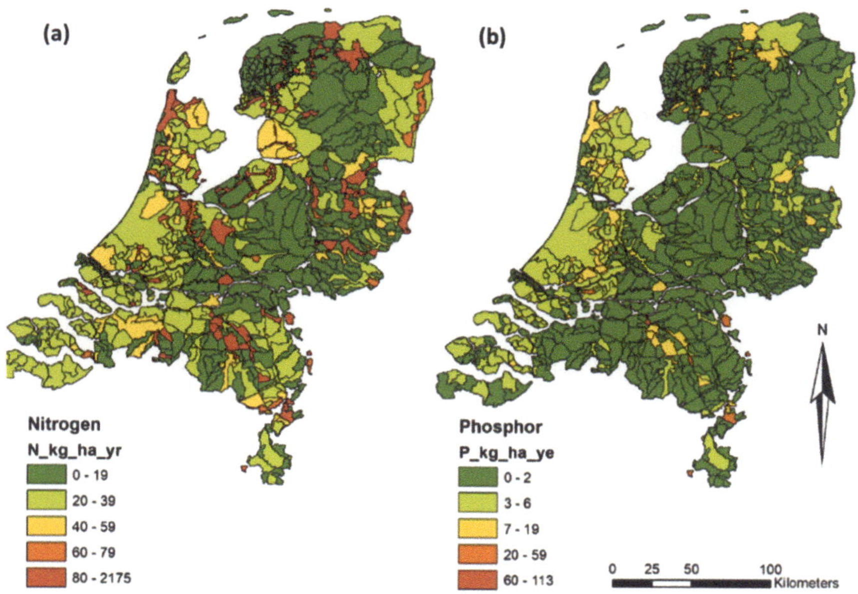

Fig. 8 Total annual (A) nitrogen and (B) phosphorus loads entering surface water bodies in catchments across the Netherlands.

relatively low phosphorus loads to their surface waters. Among the contributing sources, wastewater/stormwater treatment plants emitted large amounts of nutrients into surface waters in each catchment.

Fig. 9 shows percentage retention of phosphorus and nitrogen, i.e., the total nutrient load captured/retained in each catchment, which can be interpreted as the natural self- purification capacity of the catchments (NR service). Western parts of the Netherlands showed high percentage retention of both nutrients (darker brown in Fig. 9), while eastern parts showed low/intermediate percentage nutrient retention (lighter brown).

Fig. 10 shows the residual nutrient loads in water bodies in the catchments after natural retention. Most catchments in the southwest and south of the country had low or intermediate annual nutrient loads (green or yellow in Fig. 8) and high retention rates (brown in Fig. 9), resulting in low amounts of residual nutrients (green in Fig. 10). Some catchments with high nutrient loads (red in Fig. 8) also had low or intermediate amounts of residual nutrients (green or orange in Fig. 10). Catchments with high nutrient loads and low retention rates (mainly located in the south and east of the country) emerged as areas of interest that need further management measures to reduce the nutrient load (i.e., reduce potential demand for the NR service).

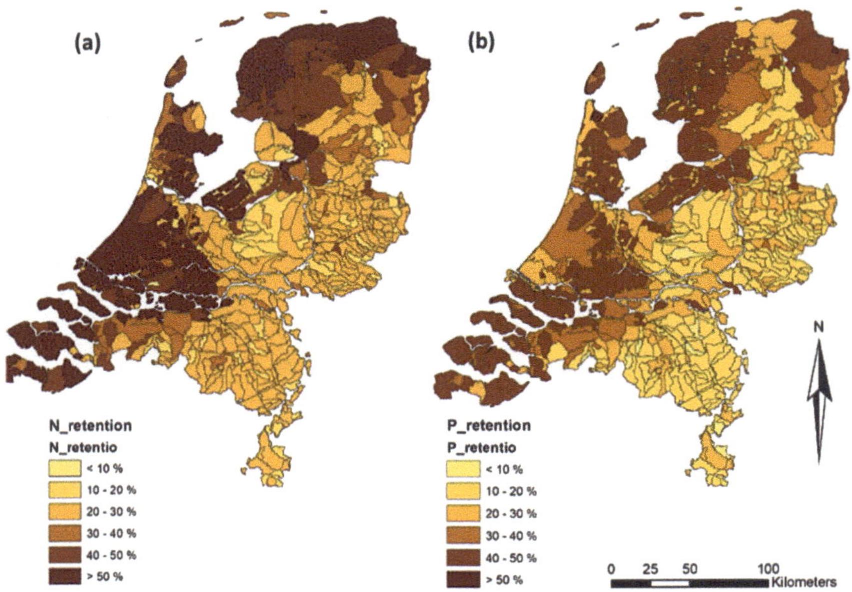

Fig. 9 Mean percentage retention of (A) nitrogen and (B) phosphorus in catchments across the Netherlands.

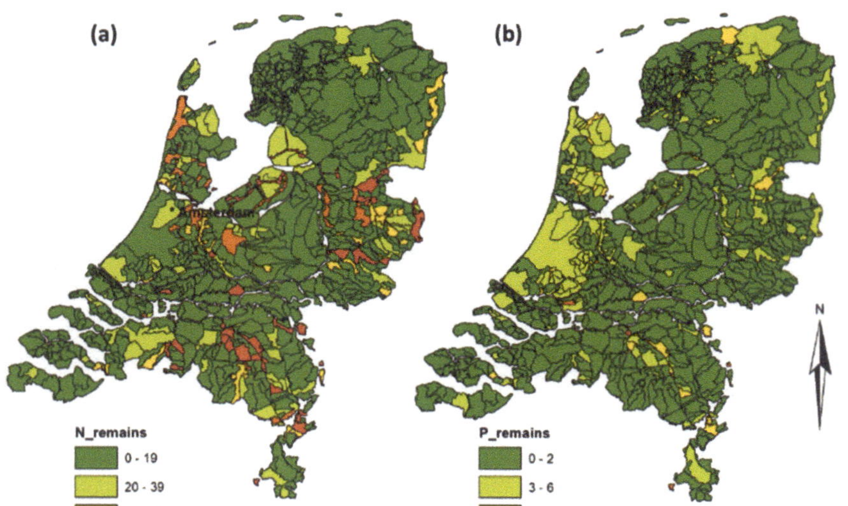

Fig. 10 Residual (A) nitrogen and (B) phosphorus loads in surface water bodies in catchments across the Netherlands after natural retention.

5. Implications for management

5.1 Lessons learned from mapping ecosystem services in the two case study areas

In the maps obtained, the city of Amsterdam exhibited a pattern of a heavily densified urban city center with a high demand for the LCR service that potentially cannot be met/supplied by natural and/or man-made ecosystems. However, open green spaces, trees and water bodies in the city can to some extent increase provision of the LCR service and potentially lower the demand for this service in nearby areas.

As most of the areas in the inner city of Amsterdam fell within the continuous urban fabric land cover class (S.L. >80%), a pattern of high demand for the LCR service emerged for those areas. However, it was difficult to detect every small patch of urban green area, including those with some impact on the supply of the LCR ecosystem service. Overall, the analysis revealed undersupply of the LCR service in and around the city of Amsterdam. This is consistent with findings in a temperature study by van der Hoeven and Wandl[41] that surface and air temperatures in Amsterdam are increasing as the city grows, reflecting the UHI effect, which can have serious impacts on the comfort and health of residents (van der Hoeven and Wandl[41]). Our study showed that demand for the LCR service to mitigate such effects remains unmet in the city.

Mapping of the NR potential supply in the Netherlands based on indicator values assigned to different land cover classes revealed a similar pattern as in maps produced by Bicking et al.[37] for the same service in northern Germany. Their study area was also dominated by low supply, with small areas of medium to high supply scattered randomly throughout. Urban, artificial areas have very limited capacity/ability for natural regulation of excess nutrients released into the environment. Forests and water systems do have this ability, but assigning those land cover classes the same indicator values, without knowing whether they have the actual capacity to provide the service or not, can lead to unrealistic results. Combining land cover maps with soil maps could be more useful in assessing the supply of the NR service. In relation to nutrient management on catchment scale, it is worth mentioning that maps do not provide sufficient information to evaluate quality changes in the recipient (e.g., water) environment. However, they can be used as a starting point for future research on the NR service in the Netherlands and can help in attributing nutrient emissions to different sources in the catchments.

5.2 The necessity of mapping ecosystem services for climate change impact mitigation and adaptation

Ecosystems provide essential services for humans, animals, and plants and are part of the solution to mitigate climate change impacts, but they are also influenced by climate change and human activities themselves.[42] Mapping ecosystem services is necessary for understanding how ecosystems contribute to human well-being and for formulating policies to protect natural resources. Ecosystem service maps are an effective tool for communicating information about land-use changes, human-induced ecosystem losses and potential for their revival, providing a visual understanding of complex human-nature interactions across a landscape. They also provide critical data for integrating ecosystem services into policy and decision-making processes.

For an accurate depiction of an ecosystem service, various aspects and criteria should be taken into account in the analysis, depending on influencing/contributing factors. For instance, mapping potential supply of the NR service based on only land cover information, without considering other processes (e.g., denitrification, water surface and bottom area, nutrient concentrations etc.), can lead to inaccurate conclusions on realization of this service. The actual capacity/ability of ecosystems also needs to be assessed and quantified for a realistic comparison of potential supply and demand for the services they provide. Using a combination of maps, data and analytical and statistical methods can help evaluate the actual net budget for a specific ecosystem service and better support identification of relevant management measures.

Assessing and mapping ecosystem services is complex due to the different processes involved, requiring more in-depth understanding and data to produce accurate results. Despite the limitations and complexities, maps of ecosystem services can assist urban planners and decision-makers in adaptive spatial planning and improve the resilience of cities to climate change impacts.

6. Conclusions

Ecosystems provide essential services for humans, animals and plants, but humans have altered ecosystems dramatically in recent decades and the pressures on ecosystems are still growing. Mapping of ecosystem services can provide better insights into the complex supply-demand system and can be a useful tool in urban development planning. Assessment of ecosystem services is also helpful in cost-benefit analyses of remediation measures, since it considers the natural capital to remediate.

Use of land cover data alone in mapping supply/demand for ecosystem services results in generalization. A range of different factors and parameters should be considered for an accurate representation of an ecosystem service. The method applied in this study is an easy-to-use tool for ecosystem service assessments and presents easy-to-interpret maps, but the results are variable. The method seems to work well for mapping the local climate regulation service, as the maps created in this study gave a clear picture of supply and demand for this service in the city of Amsterdam. In particular, applying an overlay of hexagons representing airflows linking supply and demand gave good results. The supply and demand maps created without considering the complex processes behind the service were adequate and provided information about different sources of supply in the city. When applying the same method for a much larger study area the credibility of the results diminished, as shown when mapping the nutrient regulation service in the whole of the Netherlands. The results obtained were not sufficient for assessment of this service, and better results might have been obtained if the analysis had been performed on a smaller scale and if the land cover map used had been more detailed than only general classes. In assessing the demand for the nutrient regulation service, using indicators with a range of 0–5 proved to be unsuitable and determining the nutrient loads in an area (or catchment) may be a better way of quantifying the demand for this service. It is also a way to identify so-called "hotspots," i.e., areas with high nutrient emissions that result in pollution of surface waters.

Overall, assessing and mapping ecosystem services is complex due to the many contributing processes and requires more in-depth data and inputs to achieve more accurate results. The results presented here for the city of Amsterdam can be helpful for urban planners and decision makers. They can also serve as a starting point for future assessments.

Acknowledgments

This research was supported by the Bolin Centre for Climate Research Area 7. Landscape processes and climate) at Department of Physical Geography, Stockholm University.

References

1. United Nations. *United Nations sustainable development united nations sustainable development 2015*; 2015. Available at: https://www.un.org/sustainabledevelopment/blog/2015/07/un-projects-world-population-to-reach-8-5-billion-by-2030-driven-by-growth-in-developing-countries/.

2. Kalantari Z. Enlivening our cities: towards urban sustainability and resilience. *Ambio* 2021;**2021**. https://doi.org/10.1007/s13280-021-01518-w.
3. Daily GC. *Nature's services: societal dependence on natural ecosystems*. Yale University Press; 2015.
4. Breuste J, Haase D, Elmqvist T. Urban landscapes and ecosystem services. In: Wratten S, Sandhu H, Cullen R, Costanza R, editors. *Ecosystem services in agricultural and urban landscapes*; 2013, Wiley, Chichester. p. 83–104.
5. Burkhard B, Kroll F, Nedkov S, Müller F. Mapping ecosystem service supply, demand and budgets. *Ecology* 2012;**21**:17–29.
6. Elmqvist T, Fragkias M, Goodness J, Güneralp B, Marcotullio PJ, McDonald RI, et al. *Urbanization, biodiversity and ecosystem services: challenges and opportunities—a global assessment*. Springer Nature; 2013.
7. Sirakaya A, Cliquet A, Harris J. Ecosystem services in cities: towards the international legal protection of ecosystem services in urban environments. *Ecosyst Serv* 2018;**29**:205–12.
8. Plieninger T, van der Horst D, Schleyer C, Bieling C. Sustaining ecosystem services in cultural landscapes. *Ecol Soc* 2014;**19**:59.
9. Åhlén I, Vigouroux G, Destouni G, Pietroń J, Ghajarnia N, Anaya J, et al. Hydro-climatic changes of wetlandscapes across the world. *Sci Rep* 2021;**11**:2754.
10. Kalantari Z, Ferreira CSS, Page J, Goldenberg R, Olsson J, Destouni G. Meeting sustainable development challenges in growing cities: coupled social-ecological systems modeling of land use and water changes. *J Environ Manag* 2019;**245**:471–80.
11. Reid WV. *Millennium ecosystem assessment: ecosystems and human well-being*. Island Press; 2005.
12. Elmqvist T, Setälä H, Handel SN, van der Ploeg S, Aronson J, Blignaut JN, et al. Benefits of restoring ecosystem services in urban areas. *J Curr Opin Environ Sustain* 2015;**14**:101–8.
13. Copernicus Land Monitoring Service. *Land item 2018*; 2018. Available at: https://land.copernicus.eu/pan-european/corine-land-cover/clc2018.
14. Ferreira CSS, Pereira P, Kalantari Z. Human impacts on soil. *Sci Total Environ* 2018;**644**:830–4.
15. Guerra AJT, Fullen MA, Jorge MCO, Bezerra JFR, Shokr MS. Slope processes, mass movement and soil erosion: a review. *Pedosphere* 2017;**27**:27–41.
16. Ferreira CSS, Kalantari Z, Seifollahi-Aghmiuni S, Ghajarnia N, Rahmati O, Solomun MK. Rainfall-runoff-erosion processes in urban areas. In: Rodrigo-Comino J, editor. *Precipitation*; 2021. p. 481–98.
17. Grizzetti B, Lanzanova D, Liquete C, Reynaud A, Cardoso AC. Assessing water ecosystem services for water resource management. *Environ Sci Pol* 2016;**61**:194–203.
18. Daily GC, Matson PA. Ecosystem services: from theory to implementation. *Proc Natl Acad Sci* 2008;**105**:9455–6.
19. Goldenberg R, Kalantari Z, Cvetkovic V, Mörtberg U, Deal B, Destouni G. Distinction, quantification and mapping of potential and realized supply-demand of flow-dependent ecosystem services. *Sci Total Environ* 2017;**593–594**:599–609.
20. European Environment Agency (EEA). *Urban environment*; 2010. Available at: https://www.eea.europa.eu/soer/2010.
21. Yoo J, Simonit S, Connors JP, Kinzig AP, Perrings C. The valuation of off-site ecosystem service flows: deforestation, erosion and the amenity value of lakes in Prescott, Arizona. *Ecol Econ* 2014;**97**:74–83.
22. Depietri Y, Renaud FG, Kallis G. Heat waves and floods in urban areas: a policy-oriented review of ecosystem services. *Sustain Sci* 2012;**7**:95–107.

23. Kalantari Z, Seifollahi-Aghmiuni S, von Platen HN, Gustafsson M, Rahmati O, Ferreira CSS. Using landscape connectivity to identify suitable locations for nature-based solutions to reduce flood risk. In: *The handbook of environmental chemistry*. Berlin, Heidelberg: Springer; 2021. https://doi.org/10.1007/698_2021_771.
24. Balmford A, Rodrigues ASL, Walpole M, Ten Brink P, Kettunen M, Braat L, et al. *The economics of ecosystems and biodiversity: scoping the science*. Cambridge: European Commission; 2008.
25. Bolund P, Hunhammar S. Ecosystem services in urban areas. *Ecol Econ* 1999;**29**: 293–301.
26. Bowler DE, Buyung-Ali L, Knight TM, Pullin AS. Urban greening to cool towns and cities: a systematic review of the empirical evidence. *Landsc Urban Plan* 2010;**97**:147–55.
27. Schwarz N, Bauer A, Haase D. Assessing climate impacts of planning policies—an estimation for the urban region of Leipzig (Germany). *Environ Impact Assess Rev* 2011;**31**:97–111.
28. Liu X, He B, Guo L, Huang L, Chen D. Similarities and differences in the mechanisms causing the European summer heatwaves in 2003, 2010, and 2018. *Earths Future* 2020;**8**. e2019EF001386.
29. Withers PJA, Neal C, Jarvie HP, Doody DG. Agriculture and eutrophication: where do we go from here? *Sustainability* 2014;**6**:5853–75.
30. Peel MC, Finlayson BL, McMahon TA. Updated world map of the Köppen-Geiger climate classification. *Hydrol Earth Syst Sci* 2007;**11**:1633–44.
31. Harris I, Osborn TG, Jones P, Lister D. Version 4 of the CRU TS monthly high-resolution gridded multivariate climate dataset. *Sci Data* 2020;**7**:109.
32. Boumans L, Fraters D, van Drecht G. Nitrate leaching by atmospheric N deposition to upper groundwater in the sandy regions of The Netherlands in 1990. *Environ Monit Assess* 2004;**93**:1–15.
33. van Lier HN. Historical land use changes: The Netherlands. In: Brouwer FM, Thomas AJ, Chadwick MJ, editors. *Land use changes in Europe: processes of change, environmental transformations and future patterns*. The GeoJournal Library, Dordrecht, Springer Netherlands; 1991. p. 379–401.
34. Kroes JG, Supit I. Impact analysis of drought, water excess and salinity on grass production in The Netherlands using historical and future climate data. *Agric Ecosyst Environ* 2011;**144**:370–81.
35. Central Bureau for Statistics; 2021. Available at: https://opendata.cbs.nl/statline/#/CBS/nl/dataset/03759ned/table?ts=1565190493073.
36. Burkhard B, Maes J. Mapping ecosystem services. *Adv Books* 2017;**1**, e12837.
37. Bicking S, Burkhard B, Kruse M, Müller F. Mapping of nutrient regulating ecosystem service supply and demand on different scales in Schleswig-Holstein, Germany. *One Ecosys* 2018;**3**, e22509.
38. Wolf J, Beusen AHW, Groenendijk P, Kroon T, Rötter R, van Zeijts H. The integrated modeling system STONE for calculating nutrient emissions from agriculture in the Netherlands. *Environ Model Softw* 2003;**18**:597–617.
39. Uchida R. Essential nutrients for plant growth: nutrient functions and deficiency symptoms. *Plant Nutr Manage Hawaii's Soils* 2000;**4**:31–55.
40. Hein L. Economic benefits generated by protected areas: the case of the Hoge Veluwe Forest, the Netherlands. *Ecol Soc* 2011;**16**.
41. van der Hoeven F, Wandl A. *Amsterwarm: Gebiedstypologie Warmte-Eiland Amsterdam*. vol. 4. TU Delft; 2013.
42. Potschin M, Haines-Young R, Fish R, Turner RK. *Routledge handbook of ecosystem services*. Routledge; 2016.

CHAPTER EIGHT

Groundwater in the Venetian area

Sebastiano Trevisani[a,*], Paolo Fabbri[b], and Pietro Daniel Omodeo[c]
[a]Dipartimento di Culture del Progetto, Università IUAV di Venezia, Venezia, Italy
[b]Dipartimento di Geoscienze, Università di Padova, Padova, Italy
[c]Dipartimento di Filosofia e Beni Culturali, Università Ca' Foscari, Venezia, Italy
*Corresponding author: e-mail address: strevisani@iuav.it

Contents

Abstract

Venice from the perspective of water resources and hydro-geomorphological dynamics, is emblematic of the complex geosphere-anthroposphere interactions that could characterize a coastal urban area. On the one hand, the hydrogeological setting is extremely heterogenous and complex, with interconnected processes acting at multiple spatio-temporal scales, as often happen in areas with interactions between land and marine processes. On the other hand, anthropic pressures and disturbances on the geosphere are extremely complex, heterogeneous and, again, acting at multiple spatiotemporal scales. This chapter focuses mainly on groundwater related issues on the urban area of Venice municipality, comprising the historical island and the hinterland, with some reference to the interactions with surface waters. In Section 2 of the chapter, the hydrogeological setting of the area under scrutiny is introduced, highlighting the necessity to take into account a multiscale perspective for the understanding of the hydrogeological framework. In Section 3, some historical aspects related to the hydrogeological modifications and water management are introduced, pointing out as since the Serenissima Republic, centuries ago, "landscape managers" were aware of the complexity of the geohydrological system and of the necessity to adopt an holistic approach. Finally, in Section 4, the main current issues related to groundwater in the area, including the connections with surface waters, are introduced.

Keywords: Environmental humanities, Groundwater, Heterogeneity, Hydrogeological complexity, Pollution

ISSN 2468-9289
https://doi.org/10.1016/bs.apmp.2022.10.016

1. Introduction

The management and protection of groundwater in urban areas is a growing issue worldwide, both regarding quantitative as well as qualitative issues.[1–6] In urban areas often a complex and heterogenous set of environmental stressors act on complex geological and hydrogeological settings. From this viewpoint, Venice (Figs. 1 and 2), as common to many coastal areas, is characterized by complex geosphere-anthroposphere interactions, for example regarding hydrogeological and geomorphological processes. In fact, the hydrogeological setting is heterogenous and complex, with interconnected processes acting at multiple spatiotemporal scales. On the other side of the coin, multiple anthropic processes acting at multiple spatiotemporal scales influence the geosphere.[8] First, a long record of centuries of anthropic interventions characterizes the venetian landscape[9,10] and its hydro-geomorphological setting, up to the current issues of flooding due to sea surges ("acqua alta," e.g., Ref. 11) and for environmental conservation of lagoon ecosystem, currently in a critical state due to pollution and to modifications of the hydrodynamics of the lagoon. Second, the intensity of geosphere-anthroposphere interactions increased considerably in the last century, both due to industrial development and related activities as well as due to the dense urbanization and urban sprawl[12,13] that characterize the Venetian area (Fig. 1, Bottom). In this regard, due to the hydrogeological and hydraulic framework (see Section 2) of the Venetian area, many environmental treats, such as industrial, agricultural, and urban pollution, potentially affecting its aquifers originate far from the urban area of Venice, its islands and the urban and industrial hinterland of Mestre and Marghera (Fig. 2). Accordingly, it is necessary to adopt a multiscale perspective for delineating the hydrogeological framework and the main geoenvironmental and geoengineering challenges (described in Section 4).

The historical and the scientific literature related to the Venice lagoon, its water resources, its hydro-geomorphological and ecological dynamics is extremely broad and diversified, as outlined in Sections 2–4. From this perspective, it is emblematic that the bibliographic survey conducted in 1982,[14] concerning scientific publications on the Venice lagoon, required a book of 174 pages to list the references. In the present chapter we will provide some of the most relevant references.

This chapter focuses on issues that pertain to fresh groundwater resources, with some reference to surface ones, when the interaction with

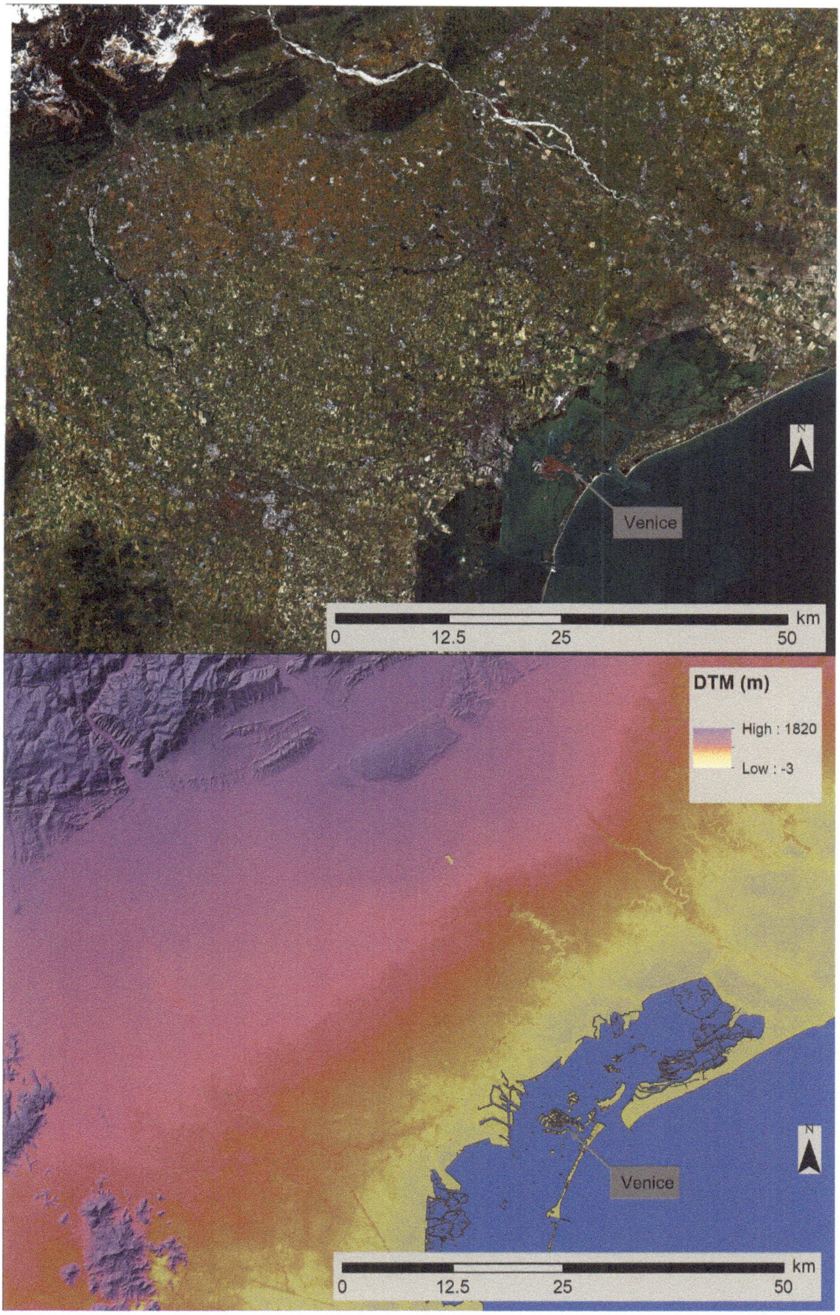

Fig. 1 (Top) Satellite imagery (Sentinel 2, ESA, 2021) covering the Veneto Region plain, on the NW margin of the Alps, and on the SE corner Venice and its lagoon; (Bottom) for the same area, the digital terrain model (DTM from Ref. 7).

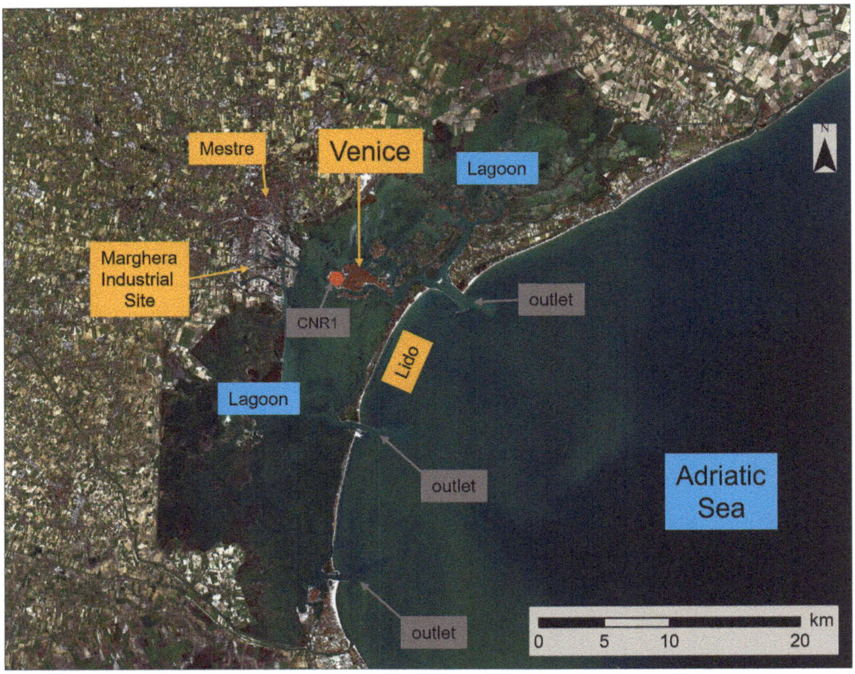

Fig. 2 Satellite imagery (Sentinel 2, ESA, 2021) covering the Lagoon of Venice, Venice historical center and its urban (Mestre) and industrial (Margera) hinterland. The location of the deep borehole "CNR1" is reported.

groundwater is relevant. The main aim is to highlight the peculiar and challenging characteristics of Venice's environment, describing some current topics that, thanks to recent studies and discoveries, are capable to shed some light on the complexities and interconnections between the multiple natural and anthropic factors characterizing this area. For example, the interactions between land surface/groundwater bodies and lagoon/marine waters represent a relevant topic, both from the perspective of freshwater quality as well as in relation to the equilibrium of the marine ecosystem (Fig. 2). Another example can be represented by the role played by the hydrogeological dynamics in the context of geomorphological equilibrium of the lagoon and of the coastal area. The chapter is composed of three main sections, in addition to the introduction (Section 1): (2) "Hydrogeological setting," describing the main hydrogeological setting of the area under scrutiny; (3) "Human-geoenvironmental interconnections in history," pointing out some historical aspects related to the hydrogeological modifications in history; (4) "Groundwater-related issues," outlining the main issues related to groundwater in the concerned area.

2. Hydrogeological setting

To understand the hydrogeological issues affecting Venice and its surroundings, it is necessary to adopt a multiscale perspective to describe its hydrogeological framework and the related interconnections between multiple geoenvironmental and anthropic processes. For example, the recharge area of the deep confined aquifers lying in the upper 300 m of sediments below Venice[15,16] is located approximately at 50 km far from the city, near the foothills of the southern Alps, in the locally referred as "high plain" area. Another example can be represented by the current sea surge hazard (locally referred as "acqua alta," e.g., Ref. 11) affecting Venice, which is related to multiple processes acting at a variety of spatio-temporal scales.

2.1 Regional hydrogeological setting

From a geostructural perspective Venice (Fig. 1) is located in a foreland basin, constrained by the SE verging frontal trusts of the Southern Alps in the north, by the NE verging trusts of the Apennines in the south and by the Dinaric belt in the east.[17,18] The plain is composed by a thick sedimentary sequence, with the depth of the bedrock increasing from the Alps, with a depth of few hundreds of meters, to the sea, with a depth of approximately 1500 m. In correspondence of Venice, the top of Eocene-Miocene marls is located roughly at 1200/1300 m below a.s.l, as deduced by the analysis of boreholes stratigraphy and seismic reflection lines.[19] The deepest borehole conducted in the historical city of Venice, borehole "CNR 1" (Fig. 2), recorded the different depositional environments and paleoclimates of the area up to depths of 900 m,[17] from the upper Pliocene to the Holocene. Near the foothills, the bedrock is much shallower, with depths less than 500 m.

From the geomorphological perspective, the plain of the Veneto Region is dominated by the alluvial megafan structures built by the alpine rivers[20,21] and fluvio-glacial processes; the coastal areas are also characterized by the presence of lagoon and marine depositional environments. The high plain, from the foothills to the area of the "springs belt" (Fig. 3) is mainly composed by coarse continental sediments, in some instances cemented, related to fluvial and fluvioglacial sedimentary processes. The low and coastal plain is characterized by a lateral and vertical alternation of finer and coarser sediments, and a deepening of the bedrock up to approximately 1500 m in the coastal area. The vertical alternation of layers of finer and coarser

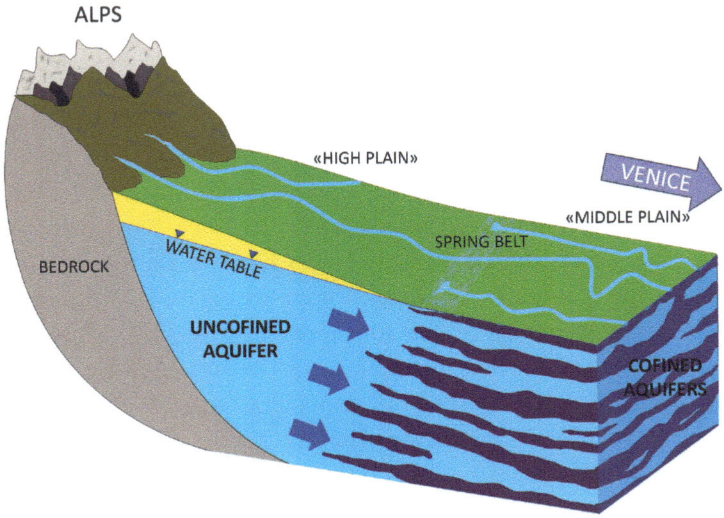

Fig. 3 Simplified hydrogeological framework of the Veneto plain: the "high plain," in the proximity of the alpine chain, represents the recharge area of an unconfined aquifer, which in turn recharges the system of confined aquifers of the "middle" and coastal plain (not presented in the scheme). In purple the confining units.

sediments constitutes a sequence of confined aquifers with a quite complex tridimensional structure and with complex connectivity patterns between aquifers.

From the hydrogeological perspective (Fig. 3), the high plain is considered an undifferentiated unconfined aquifer, with the water table intersecting the topographic surface in correspondence of the "springs belt," where in topographic local depressions the springs of some rivers are present. The waters of the unconfined aquifer recharge the complex system of superposed confined and semiconfined aquifers, characterizing the lower plain and the coastal area.[16] Unfortunately, due to the high permeability of the soil and of the sediments, and to the relatively shallow water table, the high plain unconfined aquifer is highly vulnerable to pollution. Accordingly, the multiple anthropic pressures characterizing this area[22] are a real treat for the groundwaters of the venetian plain. The risk of releasing conventional or emerging pollutants, such as industrial chemicals, hormones, and antibiotics in the recharge area[23,24] could in the future affect the deep aquifers of the Venetian area, up to now in a good qualitative state. This is also important considering the economic and strategic value of these aquifers, in terms of water quantity and quality. These aquifers are intensively exploited by public water suppliers as well by privates, including players in the rich market of mineral waters. For example, in one of the most productive areas, with an extent of 230 km^2, immediately at northeast of the Venice hinterland, covering partly the provinces of Treviso, Padova and Venice, the extraction of water from the first nine productive confined aquifers is approximately 4.6 m^3/s and 3.64 m^3/s, respectively from the private sector and from public water suppliers.[25]

In the upper 300 m of the stratigraphic sequence of the Venetian area, the sediments are representative of an alternation of shallow-marine to continental environments ranging from middle Pleistocene to present.[17] The alternation of coarse sediments (sand and less frequently gravel) and finer sediments (clay, silt, peats) constitutes a system of superposed confined and semi-confined aquifers, flowing toward the sea and with the recharge area mainly located in the Southern Alps and high plain.[26] These confined and semiconfined aquifers are relevant both for the water resources they represent as well as from the geoengineering perspective. For example, regarding water resources, a confined aquifer approximately 30 m thick and located at a depth of 280 m is highly exploited by public water suppliers and from the industry of mineral waters (e.g., https://www.sanbenedetto.it), given its productivity and high quality. Concerning geoengineering issues, the groundwater exploitation of confined aquifers from the 50s to the 70s of the 20th century was responsible for an increased subsidence rate, which practically doubled the mean average natural subsidence rate, due mainly to sedimentary cover compaction, from approximately 1 mm/y to 2 mm/y.[27] The increase in subsidence rate contributed to the increase of the sea surge hazard for Venice; for this reason measures were then taken to mitigate subsidence, such as reducing excessive groundwater pumping in the area. The interruption of excessive pumping of confined aquifers has been capable to restore natural subsidence rate, even if some studies reveal that the sealing of abandoned groundwater extraction wells, exploited by chemical industries, in some circumstances was not properly conducted, leading to some potential issues regarding water contamination.[28] Regarding the deep hydrogeological framework, it is worth mentioning the proposal[29,30] to rise Venice historical center pumping waters in the very deep confined saline aquifers, located at depths below 1000 m from sea level.

2.2 Shallow aquifers in the Venetian area

The shallow subsoil (i.e., the first 30–50 m) of the Venetian area (Fig. 2) is highly heterogeneous from the stratigraphic and hydrogeological perspective. Different locations in the area are characterized by specific active or recently active sedimentary environments: the lagoonal environment of the historical center and its islands, the coastal environment of the barrier islands (e.g., Lido island), and the fluvial environment of the hinterland. Moreover, it is characterized by a high intensity of interactions between geosphere and anthroposphere both in current as well as historical times. It is worth citing the fact that the first meters of subsoil are constituted by anthropic fillings or by natural sediments influenced by human actions, including the presence of many archeological remains from multiple ages (from Mesolithic to Modern Age), with a high frequency of Roman and medieval artifacts and infrastructures.[31–33]

The shallow stratigraphy of the historical center of Venice can be exemplified by the stratigraphic sketch of Fig. 4. The "Landfill" unit is representative of highly heterogeneous anthropic material, testifying the long and complex history of urbanization of the salt marshes and mudflats on which Venice has been built.[31,32,34] The ground level constitutes the top of the unit, with an average elevation of 1.3 m a.s.l.; the lower boundary has a highly variable depth, with an average elevation of −1.5 m a.s.l., down to depths of −6/7 m from ground level in correspondence of filled tidal channels and structures' foundations, mainly constituted by wood piles. This unit has a high heterogeneity both in terms of materials (sand, silt, clay, gravel, masonry fragments, wood tables and piles, archeological remains, etc.) as well as geotechnical characteristics. The high diversity of materials and ages, as testified by various archeological remains, is not surprising considering to the thousands of years of human presence in the area.

The "Lagoonal" unit is constituted by Holocene sediments, representative of a tidal environment, toward the hinterland influenced by freshwater inputs. The top and bottom surfaces, respectively with an average elevation of −1.5 m and −5 m a.s.l., are highly variable.[9,35] A glance of the articulated

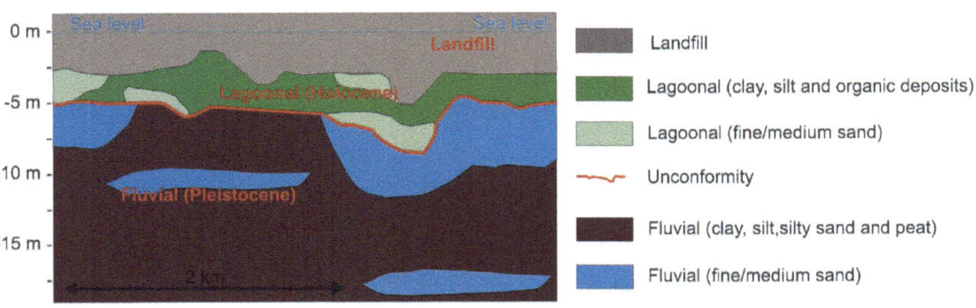

Fig. 4 Stratigraphic scheme of the shallow subsoil of the historical center of Venice. *Based on Trevisani S, Boaga J. Passive seismic prospecting in Venice historical center for impedance contrast mapping.* Environ Earth Sci *2018;* ***77****(21).*

topography of the Venetian lagoon can be obtained looking at the present complex and dynamical network of tidal channels. This unit, in correspondence of mudflats and saltmarshes, is constituted by an alternation of silts and clays, with the presence of organic matter and shells. Fine sands can be found in correspondence of scour and fill deposits of the tidal channels. The unit is characterized by high vertical and lateral lithological heterogeneity, mostly due to the dynamical characteristics of the migrating tidal channels. Some of the deepest tidal channels, in some area of the lagoon with depths up to 10–15 m, cut the underling Pleistocene sediments.

The upper part of the "Fluvial" unit is constituted by Pleistocene sediments, representative of a fluvial sedimentary environment. This unit is constituted by fine/medium sands in correspondence of fluvial channel and crevasse splays deposits, and silt and clay sediments in correspondence of distal and plug deposits. A remarkable characteristic of this unit is the presence at the top of it of a discontinuous overconsolidated silty clay level, with average thickness of 2 m. This level, locally referred as "Caranto," has been interpreted as a paleosol,[36] and marks the unconformity between Pleistocene and Holocene sediments.

In relation to this complex 3D geological architecture and to multiple anthropic disturbances, the shallow hydrogeological framework of Venice's historical center is highly heterogenous.[9] Unconfined or semiconfined aquifers can be present in the shallow subsoil of Venice, constituted by lagoonal sediments and anthropic filling. These localized aquifers are often influenced by tides and by saltwater contamination. It cannot be excluded that in some area the sand body of some ancient tidal channel could be connected with an underling Pleistocene sand body, representing potential connectivity pathways for groundwaters to emerge. Regarding the shallow confined/semiconfined aquifers of the "Pleistocene" unit, in many circumstances they are contaminated by saltwater, sometimes in relation to the deepening of lagoonal channels (see Section 4).

The barrier islands protecting the Venice lagoon, such as the Lido Island (Fig. 2), are characterized, in the shallow subsoil, by a typical coastal sedimentary environment with eolian and marine deposits; below, fluvial (Pleistocene) sediments are present. In the coastal area of Lido (Fig. 5), the main aquifer consists of sea sands with wedge-shaped geometry, increasing its thickness toward the coast and tends to shut itself off to the hinterland. This is an aquifer of remarkable lateral continuity and large dimensions (volume of about 4×10^9 m^3 and 20 m thick), with a permeability of about 10^{-5} m/s. Only the upper part of the aquifer is saturated by fresh water

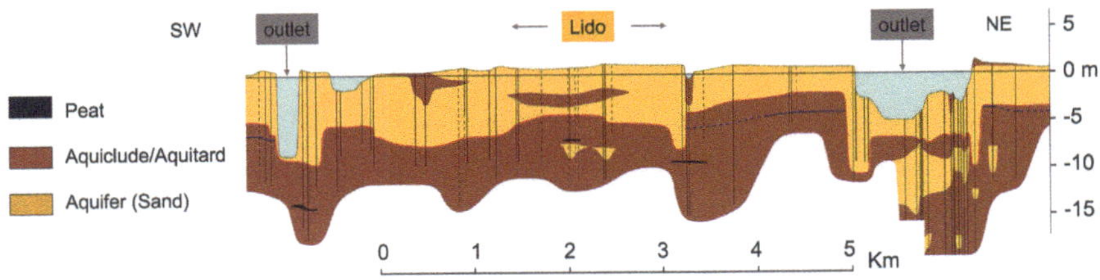

Fig. 5 Simplified shallow hydrogeological framework of the barrier island of Lido. Thin vertical lines represent borehole traces. *Modified from Fabbri P, Zangheri P, Bassan V, Fagarazzi E, Mazzucato A, Primon S, Zogno C.* Sistemi idrogeologici della provincia di Venezia. Acquiferi superficiali. Cierrre Grafica, Caselle di Sommacampagna (Verona). *288 pp. 2013.*

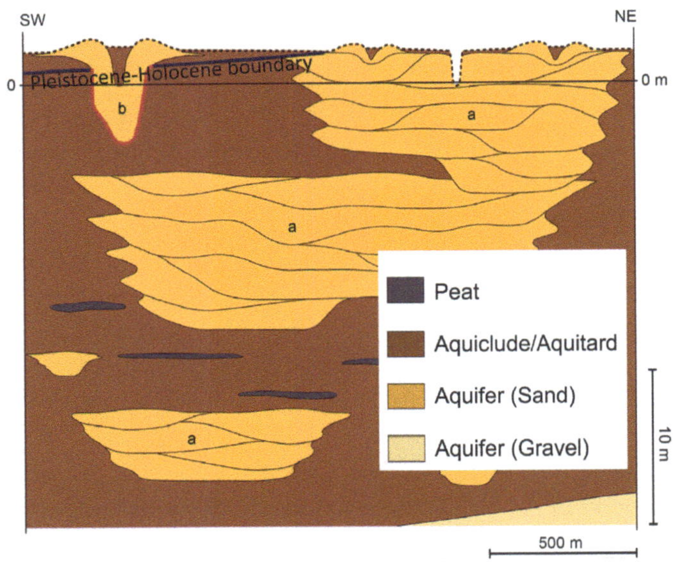

Fig. 6 Conceptual shallow hydrogeological framework for the low plain area neighboring Venice. Many aquifers are in correspondence of sand bodies (indicated with "a") of Pleistocene paleochannels; however, in some areas Holocene fluvial sands (indicated with "b") may be also present. *Modified from Fabbri P, Zangheri P, Bassan V, Fagarazzi E, Mazzucato A, Primon S, Zogno C.* Sistemi idrogeologici della provincia di Venezia. Acquiferi superficiali. Cierrre Grafica, Caselle di Sommacampagna (Verona). *288 pp. 2013.*

because the lower part is marked by saltwater intrusion. In historical times, especially in areas with large dunes, these unconfined aquifers represented relevant water reserves exploited by the city.

Moving from the lagoon to the hinterland of Mestre (Fig. 2), the transition between Pleistocene/Holocene becomes shallower, with many areas characterized by the sole presence of Pleistocene sediments (Fig. 6).

From the hydrogeological point of view, the shallow hydrogeological framework is particularly heterogenous and characterized by a complex tridimensional architecture of aquifers and with interactions with salt waters of the lagoon and of the sea. A glance of the complex and heterogenous hydrogeological framework of the shallow subsoil of Venice is possible due to the multiple studies conducted in the area. It is worth citing the geological and hydrogeological studies promoted by the former Province of Venice[16,20] capable to provide an overall and detailed overview of the area. From the quantitative perspective, a detailed and quantitative description of the shallow hydrogeological framework and dynamics has been possible only in specific areas[37–39] or for example in relation to geognostic campaigns conducted for the characterization of the polluted site of National Interest of "Porto Marghera."[40,41] Moreover, new extensive insights on the stratigraphical and hydrogeological framework of the area have been recently achieved during relevant infrastructural operations (e.g., the highway "Passante di Mestre"[16]).

These studies permitted to reconstruct the hydrogeological system, for the first 20–30 m of subsoil, on a provincial scale, shedding some light on the true complexity and heterogeneity of the area. The new picture is quite different from the one adopted by professionals and public authorities in the past, according to which the aquifer bodies, even at the technical-design level, have been frequently described as uniform tabular aquifers of low practical interest. The new picture tells a different story, with aquifers characterized by extremely heterogeneous 3D architectures and with complex connectivity patterns between the different water bodies (Fig. 7). For example, in Fig. 7, the map of the depth from ground surface of the shallow sandy aquifers (thickness between 5 and 30 m) highlights the heterogeneity and complex patterns characterizing the hydrogeological architecture. Moreover, some of the shallow aquifers have relevant water storages and fluxes,[16] posing various geoenvironmental opportunities and challenges.

In the main inland of Mestre, the predominant element is given by important sand bodies (in some cases with fine gravel at the bottom), lengthened in the NW-SE direction, 1–3 km wide and 10–15 m thick. Braided-type structures, linked to the Pleistocene fluvioglacial dynamics conditioned its geometry. Some are sub-surfacing, others have a relatively deep top (up to 10–15 m from the ground level), sometimes they are not confined or semi-confined, but most of the times confined. These aquifers have a relevant volume (total 3.5×10^9 m^3) and with a permeability of about 10^{-5} m/s; the central area is composed in the shallow subsoil, of sandy and unconfined

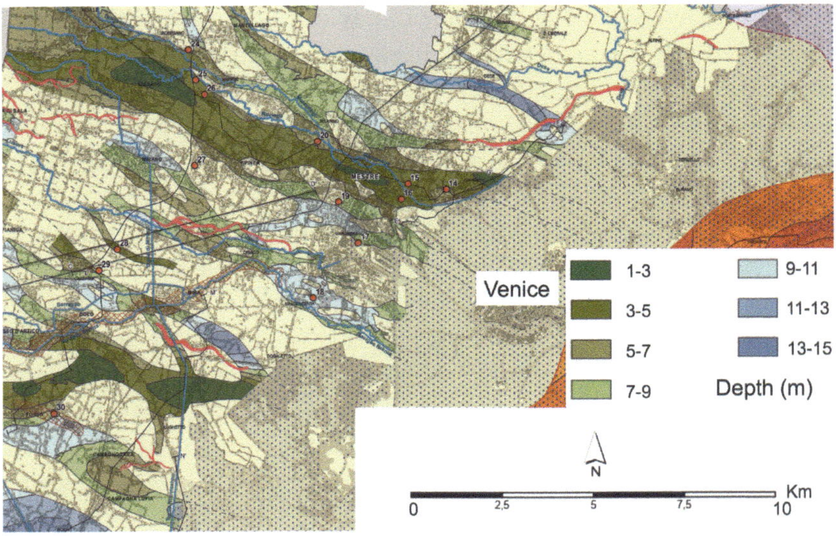

Fig. 7 Main patterns of the shallow confined/semiconfined aquifers constituted by the Pleistocene fluvial sands, flowing toward the lagoon. *Modified from Fabbri P, Zangheri P, Bassan V, Fagarazzi E, Mazzucato A, Primon S, Zogno C.* Sistemi idrogeologici della provincia di Venezia. Acquiferi superficiali. Cierrre Grafica, Caselle di Sommacampagna (Verona). *288 pp. 2013.*

aquifers, related to the main Holocene paleochannels, laterally bounded by finer clayey and silty sediments. Compared to the above-mentioned Pleistocene sandy aquifers, they are limited in volume. This is a scarcely used water reserve (domestic and sometimes for irrigation), mainly due to the poor hydrochemical quality, due to organic and inorganic pollutants, mainly nutrients. The hydrochemical quality[42–44] is influenced both by anthropic pressures (urban, industrial and agriculture) as well as in relation to the proximity to the lagoon. The interface between fresh water and salt water is relatively well known in the provincial territory between the Piave and the Adige rivers; depending on the area, it is located between 5 and 30 m from the ground level. Salinization of groundwaters can be found up to 20 km from the coastline.

The groundwater flow velocity in natural conditions is very low (e.g., cm/day); in the areas nearest to the coast and with low elevations, the groundwater tends to stagnate and flood the areas with ground level below the average sea level. The presence of a large system of hydraulic reclamation, which guarantees "clean-up zone" even in areas below the sea level, influences groundwater flow, imposing at the same time an exchange with the shallow waters. The areas below sea level occupy 622 km^2 of land, equal to about a third of the entire mainland of the venetian territory (the lagoonal area is 537 km^2). The drainage operated by the various water lifting systems[43] and by the complex hydraulic network influences the dynamics of shallow groundwaters.[41] A continuous flow between surface water and groundwater is generated from and toward the channels.

In areas near to the coast, in a range from a few hundred meters to a few kilometers, the potentiometric levels are influenced by the tidal semidiurnal cycle; in the inland areas, the influence of rainwater, infiltrating and flowing in the drainage channels, interacts with the water lifting systems. In general, the response of groundwater levels to rainfall is quite rapid (e.g., within 24 h). However, in areas with natural drainage there is a relationship between groundwater and hydrometric regime of the drainage network. Groundwater flow directions and gradients, due to the complex interconnections with surface waters dynamics, vary rapidly in time and space in relation to local factors (hydraulic network levels and flow rates, water lifting plants flow rates, tides, etc.). For example, in the areas near the lagoon margin, the groundwater hydraulic heads are influenced by the tidal semidiurnal cycle, leading to a periodic reversal of the groundwater flow directions (i.e., from and toward the lagoon).

3. Human-geoenvironmental interconnections in history

Since the Serenissima Republic, centuries ago, "landscape managers" were aware of the complexity of the geohydrological system and of the necessity to adopt an holistic approach for managing the lagoon, its waters, and natural resources. The artificial foundations of the city of Venice and its surroundings, in the middle of a lagoon that has been preserved through gigantic human efforts, marks a historically specific relation to the element of water, which has always constituted a vital resource for its inhabitants (e.g., relative to fishing, commerce, transport, and defense), but also a challenge[14,34,45,46] (Fig. 1).

For this reason, Venice still constitutes the most convenient place, both historically and symbolically, to start an inquiry into the interplay between human agency and the environment, including at a geomorphological level, according to new historiographic questions raised by ongoing debates on sustainability and climate history.[47]

As for the relevance of water for life at its most basic level, the scarcity of drinking water, considering that most of the resources were based on rainwater and rivers, in the island of Venice has been a constant problem, as is well captured by an often quoted statement of the Renaissance historian Marin Sanudo: "It [the city] is in the water but has no water" (*è in aqua et non ha aqua*).[48,49] Until the extensive exploitation of artesian aquifers was technically reliable in the nineteenth century, Venice relied on a disseminated system of cisterns, which were present all over the city in order to collect rainwater.[49] Although it cannot be excluded that some artesian water sources existed in ancient times on the islands on which today's Venice has been erected, no evident historical record of this is preserved. For this reason, the finding of a submerged source of freshwater in Cannaregio in the seventeenth century, during some excavations for the maintenance of the canals, produced much surprise and much speculation on the part of naturalists and archeologists.[50]

Environmental management and water resources protection[51] have always been at the center of the consciousness and the politics of the Venetians but it was only after the consolidation of their dominion over the mainland in the fifteenth century that the implementation of effective environmental policies and geo-engineering of the hydrographic basin of the lagoon became possible. After the submission of Padua in 1405, Venice sized control over the flow of the Brenta river in its entirety and could plan ambitious engineering interventions. Venetian aristocrats increasingly invested in the *terra ferma* (the mainland) for agricultural and animal farming purposes. This led to a process of land reclamation from the sixteenth century onward. The growing economic relevance of the mainland counterbalanced the decline of Venice's maritime supremacy and the Mediterranean commerce up to the mid eighteenth century.[52,53]

Specific institutions were created for the management of territory and resources. The most important among them was the *Magistrato alle acque*, a Magistrate for the Waters, which controlled and directed all activities related to "waters." The foundation of this permanent institution can be traced back to August 7, 1501, when a commission of *Savi alle acque* (Water Savants) was instituted. A group of knowledgeable men oversaw all matters linked with water policy in tree main areas: the lagoon, the shores and the rivers.[54] Other administrative organs integrated the functions of this magistrate. They regulated and controlled the resources, including mines (*Deputati alle miniere*) and woods, which were of strategic importance for the Venetian Arsenal. The "Superintendents to the Woods" (*Provveditori alle legna e boschi*) gave the concessions for the cutting of public and private woods, took note of the extension and quality of the reservoirs, regulated their cutting, punished the transgressors, looked over the commerce of wood and secured its flow in the capital.[55] At the crossroads of science, politics and economy, the Magistrate for the Waters had similar tasks concerning the most important resource: water. The institution was composed of savants as well as executors. Beginning in 1542 a mathematician was appointed. He gave advice to the Magistrate on issues related to engineering projects, river embankments, the harbors, and technical issues linked to water management in general. He was supported by various experts who were at his dependencies and were called *periti* or *vice-periti* depending on their position in the internal hierarchy. The activities of the Magistrate for the Waters have produced invaluable documentation throughout the centuries until its very recent suppression, in 2014. These documents are still preserved in the Venice State Archive[56] and possibly constitute a mine of historical data for geo-environmental history (on the possible use of environmental proxy data[27]).

The years of the formation of the Magistrate correspond to that age of scientific and technological development in Europe that is usually referred to as "Scientific Revolution." The problems of water management and hydrogeological engineering ignited relevant scientific debates.[57–60] These produced a large body of knowledge and highly sophisticated scientific works on topics that linked hydrology to mechanics, engineering, geology and tidal theories, to mention only the most visible disciplines.[61,62] Early publications, manuscripts and archival documents on this thematic cluster can be seen as the epistemic expression of the knowledge-mediated relationship between the water civilization of Venice and its settings.[63,64] The transformation of the territory was inextricably connected, at the level of scientific comprehension, with cosmological considerations. In fact, medieval and early-modern science considered water and earth to be elemental spheres within a cosmos made out of heavenly spheres (deputed to transport the "fixed stars," planets, sun and moon around the earth) and elemental ones (fire, air, water, earth) beneath the moon, in the framework of the scholastic reworking of Aristotelian philosophy.[65] The elementary spheres were seen as the functional parts of an organic whole within an explanatory paradigm that paved the way for the modern notion of an earth system.[66]

The specific political dimension of activities linked to water management was clearly perceived by the scientists and engineers involved in the processes of assessment and decision-making. The technical advisor (*proto*),

Cristoforo Sabbadino (1489–1560) is one of the most famous water engineers who worked for the Magistrate for the Waters. He was a strenuous defender of the preservation of the lagoon in its present state, as the most important resource of the Republic. A sonnet summarizes his scientific-political viewpoint:

You know how big your walls used to be,
Venice; now you see their present state.
And, if you do not repair its weakness,
You'll remain vulnerable, without walls.
Rivers, sea and men are your enemies.
Although You experience it, you do not believe it.
Do not wait: open your eyes and move your feet,
Otherwise you'll take a decision when it is too late.
Divert the rivers and curb men's greedy
Desires so that the sea,
Left alone, will always obey you.
If you do listen to this good advice,
The Heaven, that wisely founded you
In the waters, will donate you an eternal life.

Sabbadino, who is best known for his iconic hydrographic drawings, was involved in heated controversies over the nefarious consequences of the uncontrolled reclamation of land in the surroundings of the lagoon. Land reclamation at the expense of saltmarshes could be wished for as they permitted to develop agriculture and animal farming; however, such interventions altered the hydrodynamics and local ecological equilibrium of the lagoon. Sabbadino prescribed to reduce anthropic transformations to a minimum and strictly regulate them, with a set of prescriptions concerning land reclamation practices and water flow regulation in fish farming dedicated areas. His main opponent was the wealthy landowner, Alvise Cornaro (1489–1560), a champion of the investment into the mainland. Although many compromises would eventually be made, Sabbadino's line prevailed. In particular, he recommended to divert rivers out of the lagoon to avoid the transport of sediment into the lagoon,[64] which has constituted a major line of intervention throughout the modern period. Thus, we can regard the preservation of the insularity of Venice up to the present as Sabbadino's legacy.

Sabbadino authored works of high literary quality, which have remained in manuscript form until recent times. Among them, his *Discorsi per la laguna di Venetia* [Discourses on the Lagoon of Venice] (first printed by Cessi in 1930,[67]) are exemplary of practice-oriented science. In fact, he conceived of his activity as both practical and theoretical. As he had to deal with systemic complexity, he did not separate the natural from the anthropic. As he pointed out in his lucid analysis,

> *If one attentively considers the ruin of this lagoon produced by people, one will judge it [human activity] to be as relevant as that of rivers and the sea. Three kinds of people are responsible for this great evil: first, lords and powerful men; second, the engineers; and thirdly, individuals pursuing their own interests.*
>
> ***Sabbadino 1930: 31***

Sabbadino claimed that the negative consequences of human actions for the "hydrogeological system" are due to the short-sightedness of individuals following their specific interests, especially the unappropriated or insufficient diversion of rivers by rulers via their engineers, and the constant and deregulated activity of humans in the lagoon, such as land reclamation, fishing, and hydrodynamic modifications related to fish farms. The diversion of the Brenta into the lagoon by the Paduans, the historical enemies of Venice, caused the river's transportation of great amounts of sediments into the lagoon. Furthermore, Sabbadino chastised farmers and salt gatherers for erecting ephemeral constructions and amassing sand and materials at the shores of the lagoon. Storms and sea surges regularly destroyed their constructions during the winter and dispersed materials and sand throughout the lagoon making it less and less apt for navigation.

To sum up, geoenvironmental problems and geoengineering have always been of central concern in the management of the lagoon of Venice and its hydrological basin. The complexity of the interactions between science, politics and economy is witnessed by the creation of specific institutions devoted to the management of the resources, in particular the Magistrate for the Waters, and the emergence of a system of knowledge bridging between theory and practice and of dealing with the interactions between humans and their environment. The scientific debates that took place in the Republic of Venice reached a peak in the time of the Scientific Revolution and are particularly instructive on the difficulty to deal with a subject matter that mobilizes local knowledge and holistic cosmological-physical considerations for the sake of effective geo-environmental politics.[68]

4. Groundwater-related issues

Multiple and interrelated qualitative and quantitative issues concerning water resources characterize the Venetian area. Some issues relate to the natural characteristics of the area, e.g., the salinization of fresh water (surface and groundwater), and current climate changes, e.g., the sea level rise; however, many others are due to or enforced by anthropic factors acting at multiple scales, in space and time.[69]

A first example concerns the protection of the precious system of confined aquifers located in the first 300 m of venetian subsoil. From the quantitative point of view, the main anthropic pressures are in or in the proximity of the Venetian area, related to groundwater withdrawal from water suppliers, mineral waters corporations, and private wells (many of which unregistered and not exactly known from authorities). From the qualitative point of view, the setting is different. Some local impacts could be related to poor practices regarding well completion and casing as well as to inaccurate procedures of abandoned wells sealing[28] that can connect deep aquifers with superficial ones. However, relevant potential sources of pollution (e.g., industries, agriculture, animal farming, infrastructures, etc.) of the deep confined aquifers are mainly located far from the Venetian area (approximately at 50 km), in the recharge area of the high plain (Figs. 3 and 8). This hydrogeological area, as reported above, is highly vulnerable and is and has been characterized by strong anthropic pressures,[22,70,71] related to the

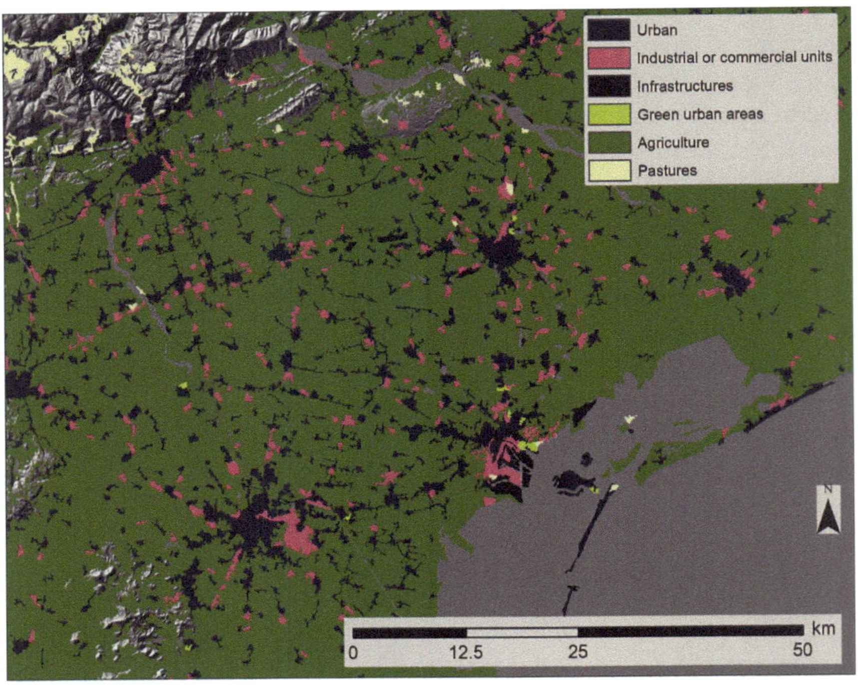

Fig. 8 Reclassified Corine land cover map (year 2006, Esa-Copernicus) highlighting the level of anthropization of the venetian landscape.

presences of urban areas, industries (e.g., metallurgy, chemicals, tanneries, etc.) infrastructures (e.g., highways), intensive animal farming and agriculture (e.g., vineyards, orchards, and other intensive productions such as soy). The land cover (Fig. 8) highlights the intensity and extensivity of anthropic pressures. It is clear from this image the level of potential treats to the water resources, even in the high plain area, very vulnerable and considered the recharge area of the confined aquifers of the low plain. An emblematic case of industrial sources of pollution (a chemical industry producing substances for stain-resistant textiles and non-stick cooking products) in the high plain is represented by the pollution with perfluoroalkyl acids (PFAS, e.g., Refs. 24,72). PFAS have been recovered in some water wells 30 km down the individual industrial source, with the beginning of the pollution processes decades ago. Moreover, the activities related to gravel extraction (for example in the Province of Treviso, covering mainly the "high plain," there are 47 quarries) in the "high plain" area represent a further treat for ground water resources, given that these generally intercepted the water table with the excavation (Fig. 9), increasing groundwater intrinsic vulnerability.

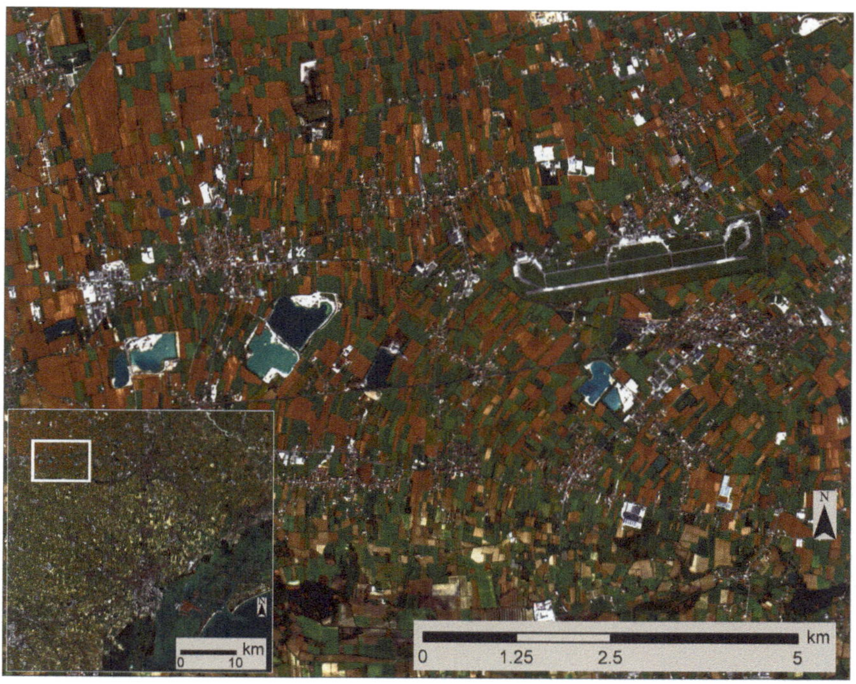

Fig. 9 A particular of the "high plain" near Treviso (Sentinel 2, ESA, 2021), highlighting the extensive quarry activities for gravel extraction intercepting the water table.

Other than this, we should consider the qualitative-quantitative impact of road-infrastructures; in this regard it is quite emblematic the new "Pedemontana Veneta" highway which is 70% undersoil and intersects this highly vulnerable area (e.g., https://www.spveneta.com/). The high plain area has a long history of industrialization and pollution; it is possible that some impacts on the confined aquifers of the middle and lower plain will be detected in the next future. Currently, these aquifers are in a good qualitative state concerning the regulatory-related contaminants monitored (e.g., according to European Directives 2000/60/CE, 2006/118/CE and 2014/80/UE). Much different it is the situation concerning the shallow semiconfined and phreatic aquifers, characterized by a general low qualitative state, for example due to ammonium and other pollutants, such as nitrates and arsenic.[43]

Focusing on Venice historical center and its hinterland, there are multiple qualitative and quantitative aspects to consider. Even in this area (Fig. 8), agriculture and animal farming represent a diffuse source of pollution[73,74] both for the shallow unconfined and semiconfined aquifers as well as for surface water bodies. This kind of pollution is quite pervasive and is particularly relevant regarding nutrient fluxes, and other pollutants, into the lagoon. This area, in addition to agriculture and urban related phenomena,[75] is subject to intense pollution in relation to the industrial site of Porto Marghera with a prevalence of chemical plants.[76] Moreover, transport infrastructures, including the area of Tessera airport at the margin of the lagoon, represent relevant pressures for the groundwater system, mainly due to deposition of dust particles, contributing to various chemicals such as polycyclic aromatic hydrocarbons and metals (e.g., Al, As, Cu, etc.).[77,78] Another treat, with unpredictable impacts, is represented by unknown pollution sources in correspondence of illegal industrial waste deposits buried in the subsoil or mixed with the filling material in infrastructures. The various findings in and near the industrial area of Marghera, with about 20 illegal waste deposits, revealed an extreme variability of the volumes and typology of substances involved.[79,80]

Both shallow groundwaters as well as surface waters have relevant interactions with anthropic factors in the urban areas. These interactions clearly include qualitative impacts related to the multiple diffuse and punctual pollution sources characteristic of urban areas.[70] However, also from the quantitative perspective interferences should be considered. For example, the high hydrogeological heterogeneity of the shallow subsoil, with aquifers mainly constituted by fluvial paleo-channels, induces a high variability in the interactions between groundwaters and surface waters. For example, this implies a high spatial variability in soil infiltration capacity, with various reflections on flood risk,[81] with a high infiltration capacity in correspondence of sand deposits (at least in conditions of relative deep-water table) and low infiltration capacity in correspondence of clay deposits. Moreover, the interaction between infrastructures and groundwater[82] can be challenging both in the phase of construction as well as in the operation conditions. The infrastructures can intersect these aquifers, modifying groundwater fluxes and also impacting the hydraulic response of the territories. An interesting example of this kind of interaction is represented by the ground water flow barrier, based on a physical barrier (sheet piles, soil bentonite slurry walls, etc.) and drainage of groundwater, designed for reducing the contamination risk toward the lagoon of the polluted site of National interest of Porto Marghera.[41,83]

The quality of groundwater, however, is not only affected by anthropic factors but also by the contribution of natural factors. For example, the shallow aquifers of the low plain can be characterized by high concentrations of metals and other elements of geological origin (e.g., arsenic, iron, manganese), frequently with concentrations above drinking water standards.[84,85] Despite the poor quality of shallow aquifers, due to the high natural background levels,[86–88] these could be exploited for irrigation purposes or in the context of heat exchange for "green building," e.g., by means of heat-pumps heat/cooling technologies.

In the coastal areas and near the margin of the lagoon, the salinization of aquifers and surface waters is another persistent issue.[89] As reported above, in the hinterland proximal to the lagoon, the hydraulic and hydrogeological setting is complex; consequently, the quantitative and qualitative modeling of the interactions between shallow aquifers, streams, channels and lagoonal waters is a difficult task.[41,88] Accordingly, a detailed mapping and forecasting of future scenarios of saltwater intrusion is difficult and site specific, even if some studies tried to map salinization vulnerability across the territory.[90]

In the lagoon, the hydrogeological setting is still more complex, with the hydrogeological qualitative and quantitative characteristics being highly heterogeneous and deeply affected by anthropic perturbations.[91] For example, the groundwater fluxes at the bottom of the lagoon are still an object of active research and many aspects has to be clarified and characterized,[26,92–94] this also in relation to the geochemical complexities in fresh water-saltwater mixing zone.[95–97] These aspects are relevant both in relation to the ecological equilibrium of the lagoon (e.g., nutrients, N/P ratio[98]) as well as for the mobilization of pollutants that can be present in the lagoon and inland sediments.[51,99–101] Another example can be represented by the quantitative and qualitative impacts of lagoonal channels excavation. The digging can intercept Pleistocene sandy aquifers, inducing a direct connection with lagoonal waters. Apart from salinization, these aquifers can be then exposed also to hydraulic influences related to the transition of mega-ships.[91] Then, in areas near the industrial sites and near freshwater inputs into the lagoon, the lagoonal sediments are often enriched in metals (e.g., Cu, Hg, Pb, Zn, see Ref. 101). The surface waters and shallow groundwaters flowing into the lagoon, heavily influenced by land reclamation pumping plants, show the presence of a wide variety of pollutants among which metals, polychlorinated biphenyls, dioxins, anionic surfactants and others that in some situations register values above the regulatory thresholds.[102] In the analysis performed in 2013 by the regional environmental agency (Arpav) the presence of various pollutants has been detected down the discharge area of some water pumping plants for land reclamation (San Giuliano and Malcontenta), including various organic compounds among which benzene and toluene.

Even in historical times, the management of freshwater resources reflected the complexity of the hydrogeological setting of Venice and surrounding Islands and salt marshes. In fact, multiple water resources[49,58] were available; these included rainwater filtered and stored by the "pozzi" (wells) hydraulic system, water resources stored in sand bodies of the coastal dunes, water transported by boats from the inland and, likely, artesian waters from shallow aquifers locally connected to the surface.

Regarding the coastal area, it is worth highlighting as the groundwater flows[89] across the barrier islands are particularly important in the relation to the sea surge Mose defense system. Moreover, the freshwater aquifers of these islands are fragile, and any perturbation of the hydrodynamic equilibrium could increase the risk of salinization, putting at risk some valuable natural protected areas located in these territories.

5. Conclusions

Despite the recent valuable efforts of public institutions and research centers, the current knowledge concerning the hydrogeological system and its environmental state of the Venetian area is relatively limited. The high complexity and dynamicity of the hydrogeological system, including the complex connectivity patterns of groundwater bodies in the low plain and coastal zones, require a much more detailed, spatiotemporally dense and coordinated hydrophysical and hydrogeochemical monitoring. The hydrogeological framework should be characterized with much more spatial detail, including the quantification of hydraulic parameters, the determination of boundary conditions and the connectivity between different water bodies. The monitoring of physical and chemical parameters should be much more pervasive in space and time. In regard to the hydrochemical quality of waters, more studies on emergent pollutants and pollutants not included in ordinary regulatory-related quality assessment practices should be promoted. Ad-hoc designed monitoring networks should be deployed for evaluating the impact of potential contamination sources and model the quantitative and qualitative state of the hydrogeological system under changing anthropic and environmental (e.g., sea level rise) forcing, as in in correspondence of new infrastructures designed in the territory. A final challenge is the capability to communicate effectively to civil society, in a clear and understandable way, the current level of knowledge regarding the groundwater resource, including the characteristics and limitations of current groundwater quality monitoring programs. Likely, in the next future, an increased public awareness could be gained adopting citizen science approaches for groundwater monitoring. These considerations even if referred to the venetian case, have a general validity, given the complex hydrogeological settings often characterizing urban areas and the limited funding provided to public agencies for environmental monitoring.

Acknowledgments

ERC Consolidator Project *EarlyModernCosmology* (Horizon 2020, GA 725883); FARE Project *EarlyGeoPraxis* (Italian Ministry of University and Research, cod. R184WNSTWH); Max Planck Partner Group *The Water City* (Max Planck Institute for the History of Science Berlin-Ca′ Foscari University of Venice).

References

1. Garg S, Motagh M, Indu J, Karanam V. Tracking hidden crisis in India's capital from space: implications of unsustainable groundwater use. *Sci Rep* 2022;**12**(1):1–16.
2. Hashimoto R, Kazama S, Hashimoto T, Oguma K, Takizawa S. Planning methods for conjunctive use of urban water resources based on quantitative water demand estimation models and groundwater regulation index in Yangon City, Myanmar. *J Clean Prod* 2022;**367**:1–11.
3. Kana JD, Ahmad Diab D, Hervé DG, N'kaya GDM, Meli'i JL, Browdon SN, et al. A combined GIS-based remote sensing and wireline log data analysis for water resource management in the economic capital district of Cameroon. *Sustain Water Resour Manag* 2022;**8**(5).

4. Aneesh TD, Nair AM, Prasad KR, Arun TJ, Srinivas R. Groundwater vulnerability assessment of an urban coastal phreatic aquifer in India using GIS-based DRASTIC model. *Groundw Sustain Dev* 2022;**19**:1–12.
5. Palma Nava A, Parker TK, Carmona Paredes RB. Challenges and experiences of managed aquifer recharge in the Mexico City metropolitan area. *Groundwater* 2022;**60** (5):675–84.
6. Sharma K, Janardhana Raju N, Singh N, Sreekesh S. Heavy metal pollution in groundwater of urban Delhi environs: pollution indices and health risk assessment. *Urban Clim* 2022;**45**:1–16.
7. Tarquini S, Nannipieri L. The 10 m-resolution TINITALY DEM as a trans-disciplinary basis for the analysis of the Italian territory: current trends and new perspectives. *Geomorphology* 2017;**281**:108–15.
8. Trevisani S, Omodeo PD. Earth scientists and sustainable development: geocomputing, new technologies, and the humanities. *Land* 2021;**10**(3):1–17.
9. Zezza F. The sedimentary structure of upper Pleistocene- Holocene deposits in Venice and its effects on the stability of the historic centre. *Rend Lincei* 2010;**21** (SUPPL. 1):211–27.
10. Trevisani S, Boaga J. Passive seismic prospecting in Venice historical center for impedance contrast mapping. *Environ Earth Sci* 2018;**77**(21):1–16.
11. Pirazzoli PA, Umgiesser G. The projected "MOSE" barriers against flooding in Venice (Italy) and the expected global sea-level rise. *J Mar Environ Eng* 2006;**8**(3):247–61.
12. Fregolent L, Tonin S. Local public spending and urban sprawl: analysis of this relationship in the veneto region of Italy. *J Urban Plann Dev* 2016;**142**(3).
13. Fregolent L, Vettoretto L. Land use regulation and social regulation: an unexplored link. Some reflections on the origins and evolution of sprawl in the Veneto "città diffusa". *Land Use Policy* 2017;**69**:149–59.
14. Pellizzato M, Scattolin M, editors. *Materiali per una bibliografia sulla laguna e sul golfo di Venezia.* Venice: Consorzio per lo sviluppo della pesca e dell'acquicoltura del Veneto; 1982. 174 pp.
15. Cultrera M, Antonelli R, Teza G, Castellaro S. A new hydrostratigraphic model of Venice area (Italy). *Environ Earth Sci* 2012;**66**(4):1021–30.
16. Fabbri P, Zangheri P, Bassan V, Fagarazzi E, Mazzucato A, Primon S, et al. *Sistemi idrogeologici della provincia di Venezia. Acquiferi superficiali. Cierre Grafica, Caselle di Sommacampagna (Verona)*; 2013. 288 pp.
17. Massari F, Rio D, Serandrei Barbero R, Asioli A, Capraro L, Fornaciari E, et al. The environment of Venice area in the past two million years. *Palaeogeogr Palaeoclimatol Palaeoecol* 2004;**202**(3–4):273–308.
18. Zecchin M, Donda F, Forlin E. Genesis of the northern Adriatic Sea (Northern Italy) since early Pliocene. *Mar Pet Geol* 2017;**79**:108–30.
19. Zecchin M, Brancolini G, Tosi L, Rizzetto F, Caffau M, Baradello L. Anatomy of the Holocene succession of the southern Venice lagoon revealed by very high-resolution seismic data. *Cont Shelf Res* 2009;**29**(10):1343–59.
20. Bondesan A, Primon S, Bassan V, Vitturi A, editors. *Le unità geologiche della provincia di Venezia. Cierre Grafica, Caselle di Sommacampagna (Verona)*; 2008. 184 pp.
21. Fontana A, Mozzi P, Bondesan A. Alluvial megafans in the venetian-Friulian plain (North-Eastern Italy): evidence of sedimentary and erosive phases during late Pleistocene and Holocene. *Quat Int* 2008;**189**(1):71–90.
22. Autorità di Bacino Distrettuale delle Alpi Orientali. *PIANO DI GESTIONE DELLE ACQUE River Basin Management Plan, Aggiornamento 2015–2021, Sintesi delle pressioni e degli impatti significativi sullo stato delle acque. Volume 3, Technical Report*; 2016. 68 pp.
23. Antonelli R, Dal Pra A. Problems of contamination and protection of underground waters in the Veneto alluvial plain (northern Italy). *Will Sci Tech* 1991;**24**(11):63–70.

24. Valsecchi S, Rusconi M, Mazzoni M, Viviano G, Pagnotta R, Zaghi C, et al. Occurrence and sources of perfluoroalkyl acids in Italian river basins. *Chemosphere* 2015;**129**:126–34.
25. Cambruzzi T, Conchetto E, Fabbri P, Zangheri P, Marcolongo E, Rosignoli A. *Risorse idriche e bilancio idrogeologico. Report, AATO, Laguna di Venezia*; 2010. 192 pp.
26. Zuppi GM, Sacchi E. Dynamic processes in the Venice region outlined by environmental isotopes. *Isot Environ Health Stud* 2004;**40**(1):35–44.
27. Camuffo D, Bertolin C, Schenal P. A novel proxy and the sea level rise in Venice, Italy, from 1350 to 2014. *Clim Chang* 2017;**143**(1–2):73–86.
28. Da Lio C, Tosi L, Zambon G, Vianello A, Baldin G, Lorenzetti G, et al. Long-term groundwater dynamics in the coastal confined aquifers of Venice (Italy). *Estuar Coast Shelf Sci* 2013;**135**:248–59.
29. Castelletto N, Ferronato M, Gambolati G, Putti M, Teatini P. Can Venice be raised by pumping water underground? A pilot project to help decide. *Water Resour Res* 2008;**44**(1):1–12.
30. Teatini P, Ferronato M, Gambolati G, Baú D, Putti M. Anthropogenic Venice uplift by seawater pumping into a heterogeneous aquifer system. *Water Resour Res* 2010; **46**(11):1–16.
31. Ninfo A, Fontana A, Mozzi P. The map of Altinum, ancestor of Venice. *Science* 2009;**325**:577.
32. Gelichi S. L'archeologia nella laguna veneziana e la nascita di una nuova città. *Reti Mediev J* 2010;**11**:2. Firenze University Press.
33. Madricardo F, Bassani M, D'Acunto G, Calandriello A, Foglini. New evidence of a Roman road in the Venice lagoon (Italy) based on high resolution seafloor reconstruction. *Sci Rep* 2021;**11**:13985.
34. Canal E. *Archeologia della laguna di Venezia 1910–2010*. Verona: Cierre edizioni; 2015.
35. McClennen CE, Ammennan AJ, Schock SG. Framework stratigraphy for the lagoon of Venice, Italy: revealed in new seismic-reflection profiles and cores. *J Coast Res* 1997; **13**(3):745–59.
36. Donnici S, Serandrei-Barbero R, Bini C, Bonardi M, Lezziero A. The caranto paleosol and its role in the early urbanization of Venice. *Geoarchaeology* 2011;**26**(4):514–43.
37. Sartore L, Fabbri P, Gaetan C. spMC: an R package for 3D lithological reconstructions based on spatial Markov chains. *Comput Geosci* 2016;**94**:40–7. https://doi.org/10.1016/j.cageo.2016.06.001.
38. Fabbri P, Gaetan C, Sartore L, Dalla LN. Subsoil reconstruction in geostatistics beyond kriging: a case study in Veneto region (NE, Italy). *Hydrology (MDPI)* 2020;**7**:15. https://doi.org/10.3390/hydrology7010015.
39. Allard D, Fabbri P, Gaetan C. Modeling and simulating depositional sequence using a latent Gaussian model. *Math Geosci* 2021;**53**(4):469–97. https://doi.org/10.1007/s11004-020-09875-0.
40. Trevisani S, Fabbri P. Geostatistical modeling of a heterogeneous site bordering the Venice lagoon, Italy. *Ground Water* 2010;**48**(4):614–23.
41. Beretta GP, Terrenghi J. Groundwater flow in the Venice lagoon and remediation of the Porto Marghera industrial area (Italy). *Hydrogeol J* 2017;**25**(3):847–61.
42. Autorità di Bacino Distrettuale delle Alpi Orientali. *PIANO DI GESTIONE DELLE ACQUE River Basin Management Plan, Aggiornamento 2015–2021, Stato e obiettivi ambientali delle acque. Volume 6, technical report*; 2016. 140 pp.
43. Arpav. *La Qualità delle acque interne in Provincia di Venezia, technical report*; 2018. 118 pp.
44. Arpav. *Qualità delle acque sotterranee, technical report*; 2019. 97 pp.
45. D'Origo W. *Venezia Origini: Fonramenti, ipotesi, metodi*. Milano: Electa; 1983.
46. Vianello R. Global climate changes in Venice lagoon: the phenomenon of "acqua Alta" and the perception of the safeguards works. *Arch Antropol Mediterr XX* 2017;**19**(2):81–96.

47. White S, Pfister C, Mauelshagen F, editors. *The Palgrave handbook of climate history*. London: Palgrave Mcmillan; 2018.
48. Sanudo M. *De origine, situ et magistratibus urbis Venetae, ovvero la città di Venetia (1493–1530)*. Milan: Cisalpino-La Goliardiaca; 1980.
49. Gentilcore D. The cistern-system of early modern Venice: technology, politics and culture in a hydraulic society. *Water Hist* 2021;**13/3**:1–31.
50. Luzzini F. Quando il Cannaregio si fece dolce. *Acque Sotterranee—Ital J Groundwater* 2019;**390**:75–7.
51. Rosati G, Solidoro C, Canu D. Mercury dynamics in a changing coastal area over industrial and postindustrial phases: lessons from the Venice lagoon. *Sci Total Environ* 2020;**743**:1–15.
52. Ventura A. Considerazioni sull'agricoltura veneta e sulla accumulazione originaria del capitale nei secoli XVI e XVII. *Stud Stor* 1968;**9**:674–722.
53. Gullino G. Quando il mercante costruì la villa: le proprietà dei Veneziani nella Terraferma. In: Cozzi G, Prodi P, editors. *Storia di Venezia*. vol 6. Roma: Treccani; 1994. p. 875–924.
54. Ventrice P. Le istituzioni e la gestione della laguna ai tempi della Serenissima. In: *Nel primo centenario del Magistrato alle Acque*. Venezia: Istituto Veneto di Scienze Lettere ed Arti; 2008.
55. Appuhn K. *A Forest on the sea: environmental expertise in renaissance Venice*. Baltimore: The Johns Hopkins University Press; 2009.
56. Mosto D, Andrea. *L'Archivio di Stato di Venezia: Indice generale, storico, descrittivo ed analitico*. vol 1. Roma: Biblioteca d'Arte Editrice; 1937.
57. Sabbadino C. In: Gobetto PGT, editor. *Il sistema Laguna a metà Cinquecento: Opere scelte pubblicate nel 450° della morte*. Mestrino (PD): Il Leggio; 2011.
58. Luzzini F. *Theory, practice, and nature in-between: Antonio Vallisneri's Primi Itineris specimen*. Berlin: Edition Open Access; 2018.
59. Maffioli CS. *Out of Galileo. The science of waters (1628–1718)*. Rotterdam: Erasmus Publishing; 1994.
60. Maffioli CS. *La via delle acque (1500–1700). Appropriazione delle arti e trasformazione delle matematiche*. Florence: Olschki; 2010.
61. AAVV. *Raccolta d'autori che trattano del moto dell'acque, 7 volumes*. Firenze: G.A. Reale; 1765–1774.
62. *Antichi scrittori d'idraulica veneta. [1919–1952]*; 1987–1988. 5 volumes.
63. Ciriacono S. *Building on water: Venice, Holland and the construction of the European landscape in early modern times*. New York and Oxford: Berghahn Books; 2006.
64. Ciriacono S. Cristoforo Sabbadino e l'idraulica europea del Cinquecento. In: *Sabbadino, Il sistema Laguna*; 2011. p. 13–28.
65. Vogel K. Das problem der relativen lage von Erd- und Wassersphäre im Mittelalter und die kosmographische revolution. In: *Mitteilungen der Österreichischen Gesellschaft für Wissenschaftsgeschichte 13*; 1993. p. 103–43.
66. Omodeo PD. *The origin of the idea of material and life cycles in the ancient cosmos of concentric spheres*; 2018. Technosphere Magazine 12, special issue "Dossier Spheres" https://technosphere-magazine.hkw.de/p/The-Origin-of-the-Idea-of-Material-and-Life-Cycles-in-the-Ancient-Cosmos-of-Concentric-Spheres-r81oYuRppVzaupFvexVXtp.
67. Cessi R. *Documenti relativi alla storia di Venezia anteriori al mille*. Padova: Gregoriana; 1942.
68. Omodeo PD, Trevisani S, Babu S. Benedetto Castelli's considerations on the lagoon of Venice: mathematical expertise and hydro-geomorphological transformations in seventeenth-century Venice. *Earth Sci Hist* 2020;**39**(2):420–46.
69. Ravera O. The lagoon of Venice: the result of both natural factors and human influence. *J Limnol* 2000;**59**(1):19–30.

70. Ostoich M, Serena F, Pozzobon A, Tomiato L. The control of small and medium sized public wastewater treatment plants in the Veneto region (North Italy): general situation, critical issues and case studies. *Water Pract Technol* 2017;**12**(4):761–79.
71. Masiol M, Giannì B, Prete M. Herbicides in river water across the northeastern Italy: occurrence and spatial patterns of glyphosate, aminomethylphosphonic acid, and glufosinate ammonium. *Environ Sci Pollut Res* 2018;**25**(24):24368–78.
72. Mastrantonio M, Bai E, Uccelli R, Cordiano V, Screpanti A, Crosignani P. Drinking water contamination from perfluoroalkyl substances (PFAS): an ecological mortality study in the Veneto region, Italy. *Eur J Pub Health* 2018;**28**(1):180–5. https://doi.org/10.1093/eurpub/ckx066.
73. Giupponi C, Rosato P. Agricultural land use changes and water quality: a case study in the watershed of the Lagoon of Venice. *Water Sci Technol* 1999;**39**(3):135–48.
74. Giupponi C, Eiselt B, Ghetti PF. A multicriteria approach for mapping risks of agricultural pollution for water resources: the Venice lagoon watershed case study. *J Environ Manag* 1999;**56**(4):259–69.
75. Bendoricchio G, Calligaro L, Carrer GM. *Consequences of diffuse pollution on the water quality of rivers in the watershed of the lagoon of Venice (Italy)*; 1999. p. 113–20.
76. Guarino C, Sciarrillo R. The effectiveness and efficiency of phytoremediation of a multicontaminated industrial site: Porto Marghera (Venice Lagoon, Italy). *Chemosphere* 2017;**183**:371–9.
77. Valotto G, Zannoni D, Rampazzo G, Visin F, Formenton G, Gasparello A. Characterization and preliminary risk assessment of road dust collected in Venice airport (Italy). *J Geochem Explor* 2018;**190**:142–53.
78. Zannoni D, Valotto G, Visin F, Rampazzo G. Sources and distribution of tracer elements in road dust: the Venice mainland case of study. *J Geochem Explor* 2016; **166**:64–72.
79. Silvestri S, Omri M. A method for the remote sensing identification of uncontrolled landfills: formulation and validation. *Int J Remote Sens* 2008;**29**(4):975–89.
80. Guerzoni S, Raccanelli S, editors. *LA LAGUNA FERITA—Uno sguardo alla diossina e agli altri inquinanti organici persistenti (POP) a Venezia*. Libreria Editrice Cafoscarina; 2003. ISBN:88-7543-014-4. 96 pp.
81. Sperotto A, Torresan S, Gallina V, Coppola E, Critto A, Marcomini A. A multi-disciplinary approach to evaluate pluvial floods risk under changing climate: the case study of the municipality of Venice (Italy). *Sci Total Environ* 2015;**562**:1031–43.
82. Bernagozzi G, Benedetti G, Continelli F, Guerra C, Briganti R, Polimeni S, et al. Impacts on groundwater flow due to the excavation of artificial railway tunnels in soils. In: *Engineering geology for society and territory—volume 6: applied geology for major engineering projects*. Springer; 2015. p. 967–70.
83. Paris A, Teatini P, Venturini S, Gambolati G, Bernstein AG. Hydrological effects of bounding the Venice (Italy) industrial harbor by a protection cutoff wall: modeling study. *J Hydrol Eng* 2010;**15**(11):882–91.
84. Carraro A, Fabbri P, Giaretta A, Peruzzo L, Tateo F, Tellini F. Arsenic anomalies in shallow venetian plain (Northeast Italy) groundwater. *Environ Earth Sci* 2013;**70**:3067–84. https://doi.org/10.1007/s12665-013-2367-2.
85. Carraro A, Fabbri P, Giaretta A, Peruzzo L, Tateo F, Tellini F. Effects of redox conditions on the control of arsenic mobility in shallow alluvial aquifers on the venetian plain (Italy). *Sci Total Environ* 2015;**532**:581–94. https://doi.org/10.1016/j.scitoten.2015.06.003.
86. Dalla LN, Fabbri P, Mason L, Piccinini L, Pola M. A local natural background level concept to improve the natural background level: a case study on the Drainage Basin of the venetian lagoon in Northeastern Italy. *Environ Earth Sci* 2018;**77**:487. https://doi.org/10.1007/s12665-018-7672-3.

87. Dalla LN, Pedretti D, Tateo F, Mason L, Piccinini L, Fabbri P. Conceptual model of arsenic mobility in the shallow alluvial aquifers near Venice (Italy) elucidated through machine learning and geochemical modeling. *Water Resour Res* 2020, 1–20;**56** (9):1–20. https://doi.org/10.1029/2019WR026234.
88. Dalla LN, Pedretti D, Casiraghi G, Abel M, Piccinini L, Fabbri P. Probability of non-exceedance of arsenic concentration in groundwater estimated using stochastic multicomponent reactive transport modeling. *Water* 2021;**13**:3086. https://doi.org/10.3390/w13213086.
89. Rapaglia J, Di Sipio E, Bokuniewicz H, Maria Zuppi G, Zaggia L, Galgaro A, et al. Groundwater connections under a barrier beach: a case study in the Venice lagoon. *Cont Shelf Res* 2010;**30**(2):119–26.
90. Da Lio C, Carol E, Kruse E, Teatini P, Tosi L. Saltwater contamination in the managed low-lying farmland of the Venice coast, Italy: an assessment of vulnerability. *Sci Total Environ* 2015;**533**:356–69.
91. Teatini P, Isotton G, Nardean S, Ferronato M, Mazzia A, Da Lio C, et al. Hydrogeological effects of dredging navigable canals through lagoon shallows. A case study in Venice. *Hydrol Earth Syst Sci* 2017;**21**(11):5627–46.
92. Rapaglia J, Ferrarin C, Zaggia L, Moore WS, Umgiesser G, Garcia-Solsona E, et al. Investigation of residence time and groundwater flux in Venice lagoon: comparing radium isotope and hydrodynamical models. *J Environ Radioact* 2010;**101**(7):571–81.
93. Viezzoli A, Tosi L, Teatini P, Silvestri S. Surface water-groundwater exchange in transitional coastal environments by airborne electromagnetics: the Venice Lagoon example. *Geophys Res Lett* 2010;**37**(1):1–6.
94. Gieskes JM, Elwany H, Rasmussen L, Han S, Rathburn A, Deheyn DD. Salinity variations in the Venice Lagoon, Italy: results from the SIOSED Project, May 2005–February 2007. *Marine Chem* 2013;**154**:77–86.
95. Critto A, Zuppi GM, Carlon C, Marcomini A. Effect of a contaminated site (The San Giuliano landfill, Venice, Italy) on the interaction between water bodies in a coastal aquifer system. *Ann Chim* 2004;**94**(4):303–14.
96. Gattacceca JC, Mayer A, Cucco A, Claude C, Radakovitch O, Vallet-Coulomb C, et al. Submarine groundwater discharge in a subsiding coastal lowland: a 226Ra and 222Rn investigation in the southern Venice lagoon. *Appl Geochem* 2011;**26**(5):907–20.
97. O'Connor AE, Luek JL, McIntosh H, Beck AJ. Geochemistry of redox-sensitive trace elements in a shallow subterranean estuary. *Mar Chem* 2015;**172**:70–81.
98. Salvetti R, Acutis M, Azzellino A, Carpani M, Giupponi C, Parati P, et al. Modelling the point and non-point nitrogen loads to the Venice lagoon (Italy): the application of water quality models to the Dese-Zero basin. *Desalination* 2008;**226**(1–3):81–8.
99. Secco T, Pellizzato F, Sfriso A, Pavoni B. The changing state of contamination in the lagoon of Venice. Part 1: organic pollutants. *Chemosphere* 2005;**58**(3):279–90.
100. Bernardello M, Secco T, Pellizzato F, Chinellato M, Sfriso A, Pavoni B. The changing state of contamination in the lagoon of Venice. Part 2: heavy metals. *Chemosphere* 2006;**64**(8):1334–45.
101. Zonta R, Botter M, Cassin D, Pini R, Scattolin M, Zaggia L. Sediment chemical contamination of a shallow water area close to the industrial zone of Porto Marghera (Venice Lagoon, Italy). *Mar Pollut Bull* 2007;**55**:529–42.
102. Regione Veneto, Arpav, Magistrato alle Acque Di Venezia, editor. *Progetto per l'integrazione delle conoscenze sui carichi inquinanti immessi nella laguna di Venezia dai bacini a scolo meccanico della gronda lagunare: rapporto tecnico. Technical report*; 2013. 85 pp.

FSC
www.fsc.org
MIX
Papier | Fördert
gute Waldnutzung
FSC® C083411

Kolibri 360 GmbH
Ferdinand-Jühlke-Straße 7
99095 Erfurt, Deutschland
produktsicherheit@kolibri360.de